Intercultural Reciprocal Learning in Chinese and Western Education

Series Editors

Michael Connelly
University of Toronto
Toronto, ON, Canada

Shijing Xu
Faculty of Education
University of Windsor
Windsor, ON, Canada

This book series grows out of the current global interest and turmoil over comparative education and its role in international competition. The specific series grows out of two ongoing educational programs which are integrated in the partnership, the University of Windsor-Southwest University Teacher Education Reciprocal Learning Program and the Shanghai-Toronto-Beijing Sister School Network. These programs provide a comprehensive educational approach ranging from preschool to teacher education programs. This framework provides a structure for a set of ongoing Canada-China research teams in school curriculum and teacher education areas. The overall aim of the Partnership program, and therefore of the proposed book series, is to draw on school and university educational programs to create a comprehensive cross-cultural knowledge base and understanding of school education, teacher education and the cultural contexts for education in China and the West.

More information about this series at
http://www.palgrave.com/gp/series/15114

Sijia Cynthia Zhu · Shu Xie · Yunpeng Ma ·
Douglas McDougall
Editors

Reciprocal Learning for Cross-Cultural Mathematics Education

A Partnership Project Between Canada and China

palgrave
macmillan

Editors
Sijia Cynthia Zhu
Ontario Institute for Studies in Education
University of Toronto
Toronto, ON, Canada

Yunpeng Ma
Faculty of Education
Northeast Normal University
Changchun City, China

Shu Xie
Faculty of Education
Northeast Normal University
Changchun City, China

Douglas McDougall
Ontario Institute for Studies in Education
University of Toronto
Toronto, ON, Canada

Intercultural Reciprocal Learning in Chinese and Western Education
ISBN 978-3-030-56837-5 ISBN 978-3-030-56838-2 (eBook)
https://doi.org/10.1007/978-3-030-56838-2

This Palgrave Macmillan imprint is published by the registered company Springer Nature Switzerland AG
The registered company address is: Gewerbestrasse 11, 6330 Cham, Switzerland

Series Editors' Foreword: Intercultural Reciprocal Learning in Chinese and Western Education

The Series and East-West Contrasting Educational Narratives

This book series focuses on Chinese and Western education for the purpose of mutual understanding and reciprocal learning between the East and the West. The East has been a puzzle for the West, romanticized or demonized depending on the times. East–West relations have a long history of inquiry, and action has often been framed in competitive, ideological, and colonialist terms. In 1926, Dewey complained that "As far as we have gone at all, we have gone in loco parentis, with advice, with instruction, with example and precept. Like a good parent we would have brought up China in the way in which she should go" (p. 188). This "paternal" attitude, as Dewey called it, has not always been so benign. Economic, cultural, and intellectual matters have often been in the forefront since the Opium Wars of the nineteenth century.

Intellectually, the East–West dynamic is equally dramatic as found in works by authors such as Said (1978), Tu Wei-ming (1993), Hall and Ames (1999), Hayhoe and Pan (2001), and many others. These writers are part of a rich conceptual knowledge across cultures literature on the historical, philosophical, cultural, and educational differences of the East and West.

Education is a vital topic of international discussion and essential component part of our global consciousness. Global discussions of economics, national and regional competition, and national and regional futures often turn to education. Meanwhile, local educational discussions take place in social environments discourse of international awareness. "How are our international neighbours doing?" "How do they teach values?" "We have to catch up." These matters are vitally important. But they are not new.

Higher education in universities and other forms of postsecondary education has occupied most of the attention. What is new, and what, in our view, is likely to have far-reaching impact, is the focus on school education and early childhood education as well as pre-service teacher education. For several reasons, not the least of which is national competition, the focus on school education has been driven by comparative achievement studies. When Shanghai school students topped the chart in the Program for International Student Assessment (PISA) studies, the information was broadcast worldwide and generated ferocious discussion. One of the positive outcomes of this discussion is comparative research interest, the process of comparing educational similarities and differences in school practices, official policies, and social cultural influences. This comparative interest is all to the good and should help frame potential positive comparative futures.

But comparative research on similarity and difference is not enough. We believe we need to reach beyond the study of similarities and differences and to explore life filled school practices of people in different cultures coming together and learning from one another. In this postmodern world of instant worldwide communication, we need to go beyond comparative premises. Ideas, thoughts, images, research, knowledge, plans, and policies are in constant interaction. This book series hopes to move our international educational research onto this collaborative and interactive educational landscape of schools, parents, communities, policy, and international trends and forces.

Series Objectives and Contribution to Knowledge

The book series grew out of our seven-year Canada–China partnership study on reciprocal educational learning between Canada and China (Xu & Connelly, 2013–2020). The partnership developed from the current global interest and turmoil over comparative education and its role in international competition. The specific series grows out of two ongoing educational programs which are integrated in the partnership, the *University of Windsor-Southwest University Teacher Education Reciprocal Learning Program* and the *Shanghai-Toronto-Beijing Sister School Network*. These programs provide a comprehensive educational approach ranging from preschool to teacher education programs.

This framework provides a structure for a set of ongoing Canada–China research teams in school curriculum and teacher education areas. The overall aim of the *Partnership* program, and therefore of the proposed book series, is to draw on school and university educational programs to create a comprehensive cross-cultural knowledge base and understanding of school education, teacher education, and the cultural contexts for education in China and the West.

The first few books in the series are direct outgrowths of our partnership study. But, because of current global conditions, there is a great deal of important related work underway throughout the world. We encourage submissions to the series and expect the series to become a home for collaborative reciprocal learning educational work between the East and the West. The starting point in our Canada–China Reciprocal Learning Partnership's is the idea of a global community in which ideas, things, and people flow between countries and cultures (Xu & Connelly, 2013).

There is intense public discussion in Canada over international relations with China. The publication of international student achievement scores that rank China at the top has resulted in growing scholarly and public discussion on the differences in our educational systems. The discussion tends to focus on economic and trade relations while educational reciprocity and reciprocal learning are often absent from educational discourse. Given that the Chinese are Canada's and Ontario's largest immigrant group and that Chinese students have statistically shown academic excellence, it is critical to explore what we can learn from Chinese philosophies of education and its educational system, and what Canada can offer China in return.

The Partnership's overall goal is to compare and contrast Canadian and Chinese education in such a way that the cultural narratives of each provide frameworks for understanding and appreciating educational similarities and differences. We expect other work generated outside our partnership grant to have different starting points and socially relevant arguments. But we do expect all series works to share the *twin goals of mutual understanding and reciprocal learning*.

Built on these twin goals, the purpose of the book series is to create and assemble the definitive collection of educational writings on the similarities, differences, and reciprocal learnings between education in the East and the West. Drawing on the work of partnership-oriented researchers throughout the world, the series is designed to:

- Build educational knowledge and understanding from a cross-cultural perspective;
- Support new approaches to research on curriculum, teaching and learning in schools and teacher education programs in response to change brought on by heightened global awareness;
- Provide a compelling theoretical frame for conceptualizing the philosophical and narrative historical trajectories of these two compelling worldviews on education, society, and culture;
- Provide state of the art reviews of the comparative Chinese and English language literature on school curriculum and teacher education;
- Model, sustainable, school to school structures and methods of communication and educational sharing between Canada, other English-speaking countries and China;
- Model, sustainable, structures and methods of initial teacher training in cross-cultural understanding;
- Contribute to a documented knowledge base of similarities, differences, comparisons, and reciprocal learnings in elementary and secondary school teaching and learning curricula.

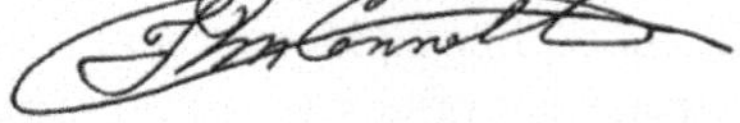

Michael Connelly
Ontario Institute for Studies in Education
University of Toronto
Toronto, ON, Canada

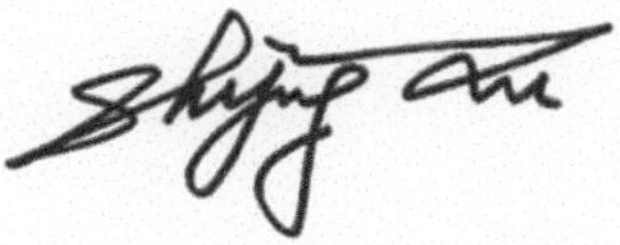

Shijing Xu
Faculty of Education
University of Windsor
Windsor, Canada

References

Dewey, J. (1926). America and the Far East. *Survey, 1 May, 1926*, 188. Later published in: John Dewey, The later works, 1925–1953, (1984). Volume 2: 1925–1927, pp. 1173–1175.

Hayhoe, R., & Pan, J. (Eds.). (2001). *Knowledge across cultures: A contribution to dialogue among civilizations*. Hong Kong: Comparative Education Research Centre, The University of Hong Kong.

Said, E. W. (1978). *Orientalism*. New York: Vintage Books.

Tu, W. (1993). *Way, learning, and politics: Essays on the Confucian intellectual*. Albany: State University of New York Press.

Xu, S., & Connelly, F. M. (Project Directors). (2013). *Reciprocal learning in teacher education and school education between Canada and China*. Social Sciences and Humanities Research Council of Canada (SSHRC) 2013–2020 [Grant 895-2012-1011].

Acknowledgments

We thank our extraordinary mathematics education research team, from professors to graduate students, whom have worked diligently over the past seven years to create a learning network that transcends cultural boundaries. From the entire mathematics team, we express our sincere gratitude to Dr. Shijing Xu and Dr. Michael Connelly for their vision, leadership, and guidance in creating the Reciprocal Learning Project. The authors are eternally grateful to all the Canadian and Chinese students, teachers, and school administrators for their participation in this project.

We would also like to acknowledge the support of the Social Sciences and Humanities Research Council of Canada Partnership Grant [Grant 895-2012-1011], the Chongqing Educational Research Institute (2015-GX-003), the University of Windsor Strategic Priority Fund, the Ontario Institute for Studies in Education (University of Toronto), and Northeast Normal University.

Sijia Cynthia Zhu
Shu Xie
Yunpeng Ma
Douglas McDougall

Introduction

As we move into a century of increased globalization, many of the science, technology, engineering, and mathematics problems that we face will require a diversity of perspectives (Medin & Lee, 2012). This means that mathematics educators will need to incorporate these diverse perspectives into the curriculum and pedagogies so that students can make connections between mathematics content knowledge and global issues in a responsible and educated manner (Orpwood, Schmidt, & Jun, 2012). In order to do so, teachers and researchers from different countries must work together to learn about each other's cultural and pedagogical practices when interacting with mathematics teaching, learning, and problem solving.

In order to increase mathematics teacher knowledge in a globalized and multicultural manner, teachers must be given more opportunities to engage in professional development that transcends the walls of their schools, districts, provinces, and countries (Timperley, Wilson, Barrar, & Fung, 2007). This is where Dr. Shijing Xu (University of Windsor) and Dr. Michael Connelly's (OISE, University of Toronto) seven-year *Reciprocal Learning in Teacher Education and School Education between Canada and China Partnership Project* fits in. This Partnership, funded by the Social Sciences and Humanities Research Council of Canada, includes six research teams of university and school educators from four Canadian and four Chinese partner institutions as well as school educators and students from more than 40 Canadian and

Chinese K-12 schools. The project was built on the foundation of the Toronto-Shanghai-Beijing Sister School Network (Connelly & Xu, 2009–2012) and the ongoing Reciprocal Learning Program developed by Dr. Xu and Dr. Shijian Chen (Southwest University, China) in partnership with the Greater Essex County District School Board (GECDSB) (Xu, 2011; Xu, Chen, & Huang, 2015) and Humanities Research Council of Canada includes six research teams of university and school educators from four Canadian and four Chinese partner institutions as well as school educators and students from more than 40 Canadian and Chinese K-12 schools. The project was built on the foundation of the Toronto-Shanghai-Beijing Sister School Network (Connelly & Xu, 2009–2012) and the ongoing Reciprocal Learning Program developed by Dr. Xu and Dr. Shijian Chen (Southwest University, China) in partnership with the Greater Essex County District School Board (GECDSB) (Xu, 2011; Xu, Chen, & Huang, 2015).

Our Mathematics Education Team is composed of university mathematics educators from two Canadian and two Chinese universities and more than a dozen of elementary school principals and teachers in Toronto and Windsor in Canada and Changchun and Chongqing in China. The partnership is founded on the idea of *reciprocal learning* (Xu & Connelly, 2013; Xu & Connelly, 2015) and "structured in such a way that Canadian and Chinese researchers, school board administrators, teachers, and students come into direct contact, creating a laboratory for the purpose of studying mutual, reciprocal learning of knowledge, values, and teaching methods" (Xu & Connelly, 2013) more than a dozen of elementary school principals and teachers in Toronto and Windsor Canada and Changchun and Chongqing in China. The partnership is founded on the idea of *reciprocal learning* (Xu & Connelly, 2013; Xu & Connelly, 2015) and "structured in such a way that Canadian and Chinese researchers, school board administrators, teachers, and students come into direct contact, creating a laboratory for the purpose of studying mutual, reciprocal learning of knowledge, values, and teaching methods" (Xu & Connelly, 2013).

By working together via a cross-cultural international reciprocal learning partnership, Chinese and Canadian teachers can learn from each other's strengths. Furthermore, they will be able to gain insight into the cultural and educational contexts that would allow for those strengths to develop and thrive (Sadler, 1979). By maintaining the learning partnership over multiple school years, the teachers will be able to develop their relationship professionally and personally, and move beyond superficial exchanges. Once the teachers have learned from each other, they will then be able to learn with each other and work together to further develop best practices for mathematics teaching and learning.

Ontario, Canada, and China are ideal partners for this collaboration as both regions are going through mathematics curriculum and teaching reforms that require new perspectives. In Canada, elementary science, math, and technology teachers are often generalists, who are in charge of teaching multiple subjects, which means that they have a lot of experience integrating different subjects. However, being generalists often means that these teachers are sometimes not as strong in subject specific content knowledge, which is important when delivering mathematics lessons (Ma, 1999).

On the other hand, Chinese elementary teachers are subject specialists and have a great deal of expertise in the specific subject that they teach (Marginson, Tytler, Freeman, & Roberts, 2013). However, this means that they have little experience integrating the science, technology, engineering, and mathematics as they are most often taught in isolated courses (Wang, Wang, Zhang, Lang, & Mayer, 1996). With the increased emphasis on relating mathematics to real-life problems, Chinese teachers will need to expand their expertise beyond mathematics and draw upon the other school subjects to create learning situations that mimic their students' lives. The complementary strengths of Canadian and Chinese elementary mathematics teachers and education systems are ideal for productive reciprocal learning. With the increased emphasis on relating mathematics to real-life problems, Chinese teachers will need to expand their expertise beyond mathematics and draw upon the other school subjects to create learning situations that mimic their students' lives. The complementary strengths of Canadian and Chinese elementary mathematics teachers and education systems are ideal for productive reciprocal learning. With the increased emphasis on relating mathematics to real-life problems, Chinese teachers will need to expand their expertise beyond mathematics and draw upon the other school subjects to create learning situations that mimic their students' lives. The complementary strengths of Canadian and Chinese elementary mathematics teachers and education systems are ideal for productive reciprocal learning. With the increased emphasis on relating mathematics to real-life problems, Chinese teachers will need to expand their expertise beyond mathematics and draw upon the other school subjects to create learning situations that mimic their students' lives. The complementary strengths of Canadian and Chinese elementary mathematics teachers and education systems are ideal for productive reciprocal learning.

It is also important for Chinese educators to increase their awareness of cultural and socioeconomic differences and their effects on teaching and learning as recent years have shown an influx of migrant workers from rural areas into industrial city centers. These migrant workers often represent China's minority cultures and have different cultural and learning needs.

When developing and proposing the Canada–China reciprocal learning partnership, Xu and Connelly (2013) point out that the research is culturally significant to Canadian classrooms because Chinese people are Canada's and Ontario's largest immigrant group. Hence, Canadian teachers must become knowledgeable about the context in which their students come from and their educational background if they hope to reach their new immigrant students. Furthermore, increased authentic experiences and awareness of China in Canada and vice versa are beneficial for teachers and students as we move toward a more globalized society. Canadian teachers must become knowledgeable about the context in which their students come from and their educational background if they hope to reach their new immigrant students. Canadian teachers must become knowledgeable about the context in which their students come from and their educational background if they hope to reach their new immigrant students.

With knowledge of those differences and similarities, the partnership strives to increase cultural understanding between the two nations and develop cross-cultural perspectives and knowledge that will lead to enhanced educational understanding in the age of globalization. This book, part of a Palgrave Macmillan's series co-edited by Drs. Connelly and Xu, showcases the reciprocal learning partnership of the mathematics research team, which is comprised of four universities: Northeast University, Changchun; Southwest University, Chongqing; University of Windsor; and University of Toronto. Over the past seven years, the University of Toronto and Northeast Normal University have been working together while the University of Windsor and Southwest University have been collaborating.

Together, the Mathematics Education Team has formed a network of university professors, research assistants, graduate students, elementary teachers and administrators, as well as schools that consistently work together on this project with a focus on mathematics education. The research and knowledge described within this book is directly supported by this network.

Sijia Cynthia Zhu
Ontario Institute for Studies in Education
University of Toronto
Toronto, ON, Canada

Shu Xie
Northeast Normal University
Changchun, China

Yunpeng Ma
Northeast Normal University
Changchun, China

Douglas McDougall
Ontario Institute for Studies in Education
University of Toronto
Toronto, ON, Canada

References

Connelly, F. M., & Xu, S. (2009–2012). *Cross-cultural narratives of schooling: Shanghai-Toronto sister school network.* Ottawa, ON, Canada: Social Sciences and Humanities Research Council (SSHRC).

Connelly, F. M., & Xu, S. J. (2019). Reciprocal learning in the partnership project: From knowing to doing in comparative research models. *Teachers and Teaching: Theory and Practice, 25*(6), 627–646.

Ma, L. (1999). *Knowing and teaching elementary mathematics: Teachers' understanding of fundamental mathematics in China and the United States.* Mahwah, NJ: Lawrence Erlbaum Associates.

Marginson, S., Tytler, R., Freeman, B., & Roberts, K. (2013). *STEM: Country comparisons.* Melbourne: Australian Council of Learned Academics.

Medin, D. L. & Lee, C. D. (2012). Diversity makes better science. Association for psychological science. Retrieved from http://www.psychologicalscience.org/publications/observer/2012/may-june-12/diversity-makes-better-science.html.

Orpwood, G., Schmidt, B., & Jun, H. (2012). *Competing in the 21st century skills race*. Canadian Council of Chief Executives.
Sadler, M. (1979). How far can we learn anything of practical value from the study of foreign systems of education? Address given at the Guildford educational conference on Saturday 20 October 1900. In J. H. Higginson (Ed.), *Selections from Michael Sadler: Studies in world citizenship* (pp. 48–51). Liverpool: Dejall & Meyorre.
Timperley, H., Wilson, A., Barrar, H., & Fung, I. (2007). *Teacher professional learning and development: Best evidence synthesis iteration wellington*. New Zealand: Ministry of Education.
Wang, W. J., Wang, J. Y., Zhang, G. Z., Lang, Y., & Mayer, V. J. (1996). Science education in the People's Republic of China. *Science Education, 80*(2), 203–222.
Xu, S. J. (2011). *Broaden the horizons: Reciprocal learning program between University of Windsor and Southwest University China*. Windsor, ON, Canada: The University of Windsor Strategic Priority Fund Proposal, University of Windsor.
Xu, S. J. (2017). *Cross-cultural schooling experiences of Chinese immigrant families: In search of home in times of transition*. Palgrave Macmillan.
Xu, S. J., Chen, S., & Huang, J. (2015). Pedagogies of working with diversity: West-East reciprocal learning in preservice teacher education. In C. Craig & L. Orland (Eds.), *International teacher education: Promising pedagogies, Volume II, working with diversity*. Emerald.
Xu, S., & Connelly, F. M. (Project Directors). (2013). *Reciprocal learning in teacher education and school education between Canada and China*. Social Sciences and Humanities Research Council of Canada (SSHRC) 2013–2020 [Grant 895-2012-1011].
Xu, S. J. & Connelly, F. M. (2015). Reciprocal learning: Comparative models and the partnership project. Keynote speech at *Reciprocal learning and symbiotic development: An international conference to promote interschool, interregional and intercultural school improvement*. Shanghai, China: East China Normal University.
Xu, S., & Connelly, F. M. (2017). Reciprocal learning between Canada and China in teacher education and school education: Partnership studies of practice in cultural context. Special issue introduction. *Frontiers of Education in China, 12*(2), 135–150.

Contents

List of Contributors

Yingying Bai School of Education, Central China Normal University, Wuhan, China

Ying Chen Ontario Institute for Studies in Education, University of Toronto, Toronto, ON, Canada

Anthony N. Ezeife University of Windsor, Windsor, ON, Canada

Jiaxin Huang Northeast Normal University, Changchun, China

Mengdie Li Southwest University, Chongqing, China

Yufeng Luo Xixin Primary School, Chengdu, China

Yunpeng Ma Faculty of Education, Northeast Normal University, Changchun, China

Douglas McDougall Ontario Institute for Studies in Education, University of Toronto, Toronto, ON, Canada

Tebatso Namanyane Southwest University, Chongqing, China

Aihui Peng Southwest University, Chongqing, China

Yanling Wang Faculty of Education, Northeast Normal University, Changchun, China

Shu Xie Faculty of Education, Northeast Normal University, Changchun, China

Bo Yu College of Teacher Education, Southwest University, Chongqing, China

Wenping Zhu Southwest University, Chongqing, China

Sijia Cynthia Zhu Ontario Institute for Studies in Education, University of Toronto, Toronto, ON, Canada

List of Figures

Chapter 3

Chapter 4

Chapter 5

Chapter 9

Chapter 10

Chapter 13

Chapter 14

Chapter 16

List of Tables

Chapter 10

Chapter 11

Chapter 12

Chapter 15

Chapter 16

PART I

Contextualizing Our Reciprocal Learning Environment in Canada & China

CHAPTER 1

General and Education Context in Ontario, Canada and Mainland China

Douglas McDougall, Yunpeng Ma, Shu Xie, and Sijia Cynthia Zhu

1 Introduction

In this chapter, we discuss the general and mathematics education context in China and the province of Ontario in Canada. There are a number of differences in the history of education in these two countries, as the

D. McDougall (✉) · S. C. Zhu
Ontario Institute for Studies in Education, University of Toronto, Toronto, ON, Canada
e-mail: doug.mcdougall@utoronto.ca

S. C. Zhu
e-mail: cyn.zhu@mail.utoronto.ca

Y. Ma · S. Xie
Northeast Normal University, Changchun City, China
e-mail: mayp@nenu.edu.cn

S. Xie
e-mail: 34474878@163.com

S. C. Zhu et al. (eds.), *Reciprocal Learning for Cross-Cultural Mathematics Education*, Intercultural Reciprocal Learning in Chinese and Western Education,
https://doi.org/10.1007/978-3-030-56838-2_1

educations systems have developed through a variety of twists and turns. We will first discuss the background of basic education reform and development in China. Then we will discuss the development of education in Canada and the province of Ontario.

2 The Background of China's Basic Education Reform and Development

Basic education was a foundational project for improving the national quality of life in China (Zhu, 2018). Development of education in China was one of the nation's top priorities. Since the founding of the People's Republic of China, basic education reform has been shaped from experience, as well as the continuous exploration of developing educational ideas through changes in society and technology.

The development of basic education in China can be divided into four stages. The first stage was the period between 1946 and 1956. During this period of national economic recovery and socialist transformation in China, basic education reform consisted of two tasks: reforming education and creating a new educational theoretical system. In 1951, the Central People's Government Administration Council created a document titled *Decision on Reforming the Academic System*, which clearly stipulated the new academic system of the People's Republic of China.

The second stage is the adjustment period of education reform (1956–1976). In 1957, Chairman Mao (1992) proposed the first education policy after the founding of the People's Republic of China, so that the educated could develop through moral, intellectual, and physical education as well as being conscious socialists and knowledgeable workers. The third stage was the comprehensive recovery period of reform and the nation's opening (1976–1999). In 1976, the Ministry of Education stipulated the *Provisional Draft of a Full-Time Ten-Year Elementary and Middle School Teaching Plan*, which uniformly stipulated that the full-time elementary and middle school education system should be ten years—five years for elementary schools and five years for middle schools.

In 1985, the Ministry of Education's *Decision of the Central Committee of the Communist Party of China on the Reform of the Education System* proposed the implementation of nine-year compulsory education, a system of operating schools at the central, provincial, and central city levels, and schools implementing a principal responsibility system. In 1986, the National People's Congress created the *Compulsory Education*

Law of the People's Republic of China, which aligned China's basic education with the legal system. In 1993, the Ministry of Education's *Outline of China's Educational Reform and Development* stipulated that compulsory education should be basically universalized. In 1999, the Ministry of Education's *Decision of the Central Committee of the Communist Party of China and the State Council on Deepening Education Reform and Comprehensively Promoting Quality Education* stipulated that the basic popularization of compulsory education and the elimination of illiteracy were the foundation for comprehensively promoting quality education.

The fourth, and current, stage is to deepen the reform of basic education (2001–present). In the twenty-first century, with the establishment of a knowledge-led economy and information-oriented society, China issued the *National Medium and Long-term Education Reform and Development Plan* with the goal of adapting to the general trend of education development and to provide a plan for the balanced development of compulsory education (Ministry of Education, 2000). In 2001, China's Ministry of Education issued the *Decision on the Reform and Development of Basic Education* and carried out a new round of curriculum reform. Aimed at the strategic position of China's basic education in the country, it proposed to deepen education reform and solidly promote quality education and thus accelerate the construction of a new basic education curriculum system that meets the requirements of quality education.

In 2011, China proposed to amend the compulsory education curriculum standard—"*Full-time Compulsory Education Curriculum Standard*" (2011 Edition), which was developed from the "*Full-time Compulsory Education Curriculum Standard (Experimental Draft).*" The structure, ideology, curriculum goals, content standards, and implementation suggestions had been modified to highlight the cultivation of students' innovative awareness. In 2019, China proposed the *Opinions on Deepening the Quality of Compulsory Education and Teaching* to implement the spirit of the Nineteenth National Congress of the Communist Party of China and the deployment of the National Education Conference, to accelerate the modernization of education, and to provide a satisfactory education for the people.

2.1 The Main Elements of Basic Education

Basic education currently includes early childhood education, primary education, and general secondary education (Wu & Peng, 2016). "Basic education implemented a system of local responsibility and hierarchical management under national guidance" (Liu, Peng, & Luo, 2020, p. 51) under local guidance. The structure of the school system was mainly 6-3-3. Ordinary elementary schools were for six years, and the ordinary middle schools were divided into two stages of three years called primary and advanced. There are still a few areas in the country that implement basic education as 5-4-3. In this implementation model, ordinary primary school was a five-year system, and ordinary middle schools were still separated into the primary and advanced stages, which were four-year and three-year, respectively.

Compulsory education occurs through grades 1–9, and enrollment in school begins at the age of six. Elementary school teaching is divided into subjects, and teachers teach one subject. However, in some areas, there are small-sized classes and the basic education work unit is composed of one to three teachers in each class, and the number of students is around 30. In this case, one teacher teaches all subjects in the elementary grades. Teachers are "all-round" (always available) to undertake the teaching activities of a class, and the routine management of classroom affairs. In some rural areas, due to the shortage of teacher resources, a mixed-class teaching was adopted, meaning multiple grades in one class. In these areas, the ordinary junior high school and ordinary high school still maintain subject-specific teachers.

3 The Background of Education in Canada and Ontario

Canada is very multicultural country home to 37.8 million people, living in ten provinces and three territories. Ontario is the largest province in Canada with about 14.7 million people (Statistics Canada, 2011). It is located between the provinces of Manitoba and Quebec, and most of the province is located north of the United States. Before being colonized by the British and French in the 1600s, Ontario was home to aboriginal peoples and the province lies mostly on treaty land.

There are over 2 million children from age 5 to 18 in Ontario's K to 12 schools (Education Facts, 2018–2019). There are public schools, Catholic public schools, in English and French in Ontario where curriculum and educational policy are the responsibility of the provinces and territories. There is no Canadian Department or Ministry of Education. At age 4, students attend junior kindergarten and at age 5, senior kindergarten.

Then they start in grade one at age 6. Elementary school is K to grade 8 and secondary school is grades 9 to 12.

Some elementary schools are separated into two separate schools: elementary school and middle school. The middle school is usually grades six to eight or grades seven to nine. A common model for elementary school is K to grades 5 or K to grades 6. The designation of elementary school and middle school differs across school districts and across provinces.

There are seven subjects in the Ontario elementary curriculum. In most schools, these subjects are language, mathematics, science and technology, social studies (history and geography), health and physical education, arts, and French as a second language. It is common that the elementary school teacher to teach all subjects except French and perhaps some of the arts-based subjects (visual arts, drama, and music). Some middle schools use a moderated secondary school system where a teacher may teach fewer subjects but more classes of the subject. These specialized teachers teach other classes than their own.

Secondary school starts in grade 9 in Ontario. The students are placed into one of three possible streams for some courses including mathematics. The streams are academic, applied and locally developed course. The students then follow these streams through to university for students in the academic stream, and college for students in the applied and locally developed course streams. There is discussion in school districts to merge the students in grade 9 into one stream. This process is called destreaming.

There are a variety of assessment methods used in elementary, middle, and secondary schools. The teacher does the assessment and then a report card is sent home at least two times per year. The report cards summarize the students' performance and information about learning skills and work habits. Assessment of elementary students is completed in schools by teachers and is done on a continuous basis and communicated to parents in the form of report cards three times a year.

There are large-scale provincial assessment tests written each year in grades 3 and 6. These tests focus on mathematics, writing, and reading and are administered at the school by the classroom teacher. The tests are marked and analyzed by a provincial organization called the Education Quality and Accountability Office (EQAO). These EQAO scores are submitted to the school and placed in the students' academic record. They are intended to provide an understanding of the performance of the grade 3 and 6 students over their elementary school career and to assist teachers and principals in identifying areas for improvement. The tests are administered in May each year.

3.1 *History of Mathematics Education in Ontario*

Mathematics in North America has a long history. As early as 1871, there has been a tension between science and mathematics and where they should fit into the curriculum. Mathematics was seen as a tool to support science but it was not a field of study (Bing, 2015). The Ontario provincial department of education was exploring the possibility that mathematics be treated as separate subject similar to reading and writing.

In the early 1900s, there were complaints by senior educators that there was an overemphasis on mathematics and that there should be a greater emphasis on the sciences. Over that short period of time, mathematics went from being absent from the curriculum to being overemphasized. Over these many years, the place of mathematics in the Ontario curriculum has been discussed and transformed without settling on how important mathematics should be to the citizens on the province.

In the middle of the twentieth century, a greater emphasis was placed on mathematics. The Cold War and the race to the moon increased the interest in mathematics (Bing, 2015; Cole, 2013). About the same time, the "Hall-Dennis" (Ontario Department of Education, 1968) report on Ontario education focused on identifying the purpose of education. The report suggested that mathematics should not be necessarily a subject but a tool to support a curriculum focused on communication. They also suggested that Ontario should have a curriculum that emphasizes concepts rather than subjects. However, they did conclude that mathematics as a subject was still important and that every student should be well versed in the subject (Memon, 2006).

Over the next 30 years, a number of task forces, commissions, and reports were written to improve education in Ontario. Some reports indicated that learning mathematics should be more comprehensive than understanding a set of facts and basic skills and proposed that applications of mathematics be situated in science, technology, and business (Ontario, 1994). This report also encouraged meta-cognitive strategies. More significantly, it proposed that destreaming of mathematics at the secondary school level should occur. This is particularly significant because mathematics at the secondary level has been streamed for many years.

In 1993, a commission proposed the establishment of the EQAO (Education Quality and Accountability Office, 2017). The EQAO was formed to provide accountability to parents, educators, and students about student performance in reading, writing, and mathematics. The tests were to be administered at grades 3, 6, and 9 and the results were to be published for the schools and the community. These large-scale provincial assessment tests provide statistical information to schools and parents about students' performance. While these tests have been controversial, we will not describe the benefits and challenges of these tests in this book. In 2005, the Ontario province released the current version of the mathematics curriculum. In Chapter 3 of this book, we explore the Ontario curriculum in more detail.

It is clear to us that mathematics has had a varied history in Ontario. There is a cycle between mathematics being a tool for science, technology, and business, to being an exploratory subject with its own field to a call to "back to the basics." There is a call to better explore what we should be teaching. Should we be focusing on knowledge, application, using a set of tools, on calculation and computation, and/or a combination of all of these possibilities? As we learn more about other countries, particularly China, we might have a better sense of the purpose and value of mathematics.

4 International Large-Scale Exams in Canada and China

International assessment studies such as the Trends in International Mathematics and Science Study (TIMSS) and Programme for International Student Assessment (PISA) have also compared the mathematics teaching and achievement of different nations. TIMSS, which is conducted by the International Association for the Evaluation of Education Achievement, is an assessment given to fourth and eighth grade students every four years that evaluates their proficiency in mathematics and science. In addition to content-based assessment, TIMSS also collects data from students, teachers, and school principals on factors related to educational opportunities in mathematics and science (Niss, Emanuelsson, & Nystrom, 2013).

As part of the TIMSS assessments, there have also been studies such as the 1999 video study that collected and analyzed over 1000 videos of in class teaching from teachers in eight countries. These videos were later compiled with transcripts and related materials and made publically available through an online database. PISA is conducted by the Organization for Economic Co-operation and Development every three years that surveys 15-year-olds in mathematics, science and reading literacy, and cross-curricular competencies such as learning strategies (OECD, 2017). In 2015, 72 countries/regions participated in PISA.

According to Cai, Mok, Reddy, and Stacey (2016), these large-scale assessments can be beneficial for nations that are hoping to monitor the progress of new initiatives, or for comparing their own achievement in respects to similar nations. Furthermore, for developing nations, these assessments provide a model for setting up their own method of monitoring the effects of policy and curriculum changes (Cai et al., 2016). In addition to the data and analysis done by OECD, IEA, and their partner agencies, TIMSS and PISA have also catalyzed further comparative studies and inquiries into what top scoring countries are doing that lead to students' success (Munson, 2011; Paine & Schleicher, 2011).

However, Feniger and Lefstein (2014) and Tsai and Li (2017) caution educators on their interpretation and use of large-scale international assessment results as they often neglect cultural and contextual influence, and function on the assumption that achievement is a result of national structures and policies. Furthermore, Tsai and Li (2017) suggest that tests such as TIMSS exacerbate the "teaching to the test culture in mathematics teaching and learning" (p. 1264). Niss et al. (2012) state that, in order for TIMSS and PISA to be "meaningful and reasonable in participating countries," there must be a "fair amount of harmonization of items, item types, response formats and score coding" (p. 1000). In this way, the mathematics is not necessarily embedded within the appropriate cultural, technological, or socioeconomic conditions, which directly influence educational performance (Feniger & Lefstein, 2014).

The international large-scale exams are important to study to better understand the similarities and differences in Canadian and Chinese mathematics education. China has consistently ranked high on these exams while Canada has been in the middle of these results. The next chapter will explore the historical origins of mathematics education in China and Canada.

References

Bing, A. C. (2015). *Pedagogical progressivism and science, technology, engineering, and math education: A shared historical landscape in Ontario, 1871–1971* (MA thesis). University of Toronto, Canada.

Cai, J., Mok, I. A. C., Reddy, V., & Stacey, K. (2016). *International comparative studies in mathematics: Lessons for improving students' learning*. SpringerOpen.

Cole, J. (2013). Alpha children wear grey: Postwar Ontario and Soviet education reform. *Historical Studies in Education/Revue d'histoire de l'éducation, 25*(1), 55–72. https://doi.org/10.32316/hse/rhe.v25i1.4300.

Education Quality and Accountability Office (EQAO). (2017). *Highlights of the provincial achievement results, 2016–2017.* Retrieved from: http://www.eqao.com/en/assessments/results/communication-docs/provincial-report-highlights-elementary-2017.pdf.

Emanuelsson, J., Niss, M., & Nyström, P. (2012). Methods for studying mathematics teaching and learning internationally. In M. Clements, A. Bishop, C. Keitel, J. Kilpatrick, & F. Leung (Eds), *Third international handbook of mathematics education. Springer International Handbooks of Education*, vol. 27. New York, NY: Springer.

Feniger, Y., & Lefstein, A. (2014). How not to reason with PISA data: An ironic investigation. *Journal of Education Policy, 29*(6), 845–855.

Government of Ontario. (1994). *For the love of learning*. Retrieved June 22, 2019, from: http://www.edu.gov.on.ca/eng/general/abcs/rcom/main.html.

Liu, J., Peng, P., & Luo, L. (2020). The relation between family socioeconomic status and academic achievement in China: A meta-analysis. *Education Psychological Review, 32*, 49–76.

Mao, Z. (1992). *Comrade mao zedong on education*. Beijing: People's Education Press.

Memon, N. A. (2006). *Contextualizing Hall-Dennis: The rise of progressive educational practices in Ontario, 1968–1972* (MA thesis). University of Toronto, Canada.

Ministry of Education of the People's Republic of China. (1976). *Provisional draft of a full-time ten-year elementary and middle school teaching plan.* Retrieved from: http://www.moe.gov.cn/.

Ministry of Education of the People's Republic of China. (1985). *Decision of the Central Committee of the Communist Party of China on the Reform of the Education System*. Retrieved from: http://www.moe.gov.cn/.

Ministry of Education of the People's Republic of China. (1986). *Compulsory education law of the People's Republic of China*. Retrieved from: http://www.moe.gov.cn/.

Ministry of Education of the People's Republic of China. (1993). *Outline of China's educational reform and development.* Retrieved from: http://www.moe.gov.cn/.

Ministry of Education of the People's Republic of China. (1999). *Decision of the Central Committee of the Communist Party of China and the State Council on deepening education reform and comprehensively promoting quality education.* Retrieved from: http://www.moe.gov.cn/.

Ministry of Education of the People's Republic of China. (2000). *National medium and long-term education reform and development plan.* Retrieved from: http://www.moe.gov.cn/.

Ministry of Education of the People's Republic of China. (2001a). *Decision on the reform and development of basic education.* Retrieved from: http://www.moe.gov.cn/.

Ministry of Education of the People's Republic of China. (2001b). *Full-time compulsory education curriculum standard (Experimental draft).* Retrieved from: http://www.moe.gov.cn/.

Ministry of Education of the People's Republic of China. (2005). *Compulsory education management system.* Retrieved from: http://www.moe.gov.cn/.

Ministry of Education of the People's Republic of China. (2011). *Full-time compulsory education curriculum standard* (2011th ed.). Beijing: Beijing Normal University Press.

Munson, L. (2011). What students really need to learn. *Educational Leadership, 68*(6), 10–14.

Niss, M., Emanuelsson, J., & Nyström, P. (2013). Methods for studying mathematics teaching and learning internationally. In M. A. (Ken) Clements, A. J. Bishop, C. Keitel, J. Kilpatrick, & F. K. S. Leung (Eds.), *Third international handbook of mathematics education* (pp. 975–1008). New York: Springer.

Ontario Department of Education. (1968). *Living and learning: The report of the provincial committee on aims and objectives of education in the schools of Ontario.* Toronto, ON: Ontario Department of Education.

Organization for Economic Co-operation and Development. (2017). *OECD science, technology and industry scoreboard 2017: The digital transformation.* Paris: OECD Publishing.

Paine, S. L., & Schleicher, A. (2011). *What the U.S. can learn from the world's most successful education reform efforts.* New York: McGraw-Hill Research Foundation.

People's Government Administration Council. (1951, October 3). Central People's Government Administration Council reforming the academic system. *People's daily.*

Statistics Canada. (2011). *Immigration and ethnocultural diversity in Canada.* Retrieved from: http://www12.statcan.gc.ca/nhs-enm/2011/as-sa/99-010-x/99-010-x2011001-eng.pdf.

Tsai, T.-L., & Li, H.-C. (2017). International comparative studies in mathematics education: Are we obsessed with the international rankings of measured educational outcomes? *International Journal of Mathematical Education in Science and Technology, 48*(8), 1262–1267.

Wu, J., & Peng, Z. (2016). The reform of the basic education system in New China: History, experience and prospects. *Education and Teaching Research*, 10.

Zhu, Y. (2018). Achievements and experiences of China's educational reform in the past 40 years. *Educator, 32,* 11–15.

CHAPTER 2

Mathematics Teacher Education in Ontario, Canada and Mainland China

Yunpeng Ma, Douglas McDougall, and Shu Xie

1 Overview of Mathematics Teacher Education

With the development of the global economy and education today, we all deeply understand the importance and necessity of improving the quality of education, and the outstanding level of teachers is an important guarantee for the quality of education. In this chapter, we discuss the pre-service and in-service mathematics teacher education in China and Ontario, Canada. The identification of teacher education candidates and the road to teacher certification have some similarities but also some very

Y. Ma · S. Xie
Northeast Normal University, Changchun City, China
e-mail: mayp@nenu.edu.cn

S. Xie
e-mail: 34474878@163.com

D. McDougall (✉)
Ontario Institute for Studies in Education, University of Toronto, Toronto, ON, Canada
e-mail: doug.mcdougall@utoronto.ca

S. C. Zhu et al. (eds.), *Reciprocal Learning for Cross-Cultural Mathematics Education*, Intercultural Reciprocal Learning in Chinese and Western Education,
https://doi.org/10.1007/978-3-030-56838-2_2

different expectations. Teacher education plays a vital role in cultivating teachers and is an important support for education. At the same time, it has a fundamental and important strategic position in the development of provincial and national education.

2 Pre-Service Mathematics Teacher Education in China

The selection and employment of mathematics teachers in primary and secondary schools is carried out through a series of strict standards and assessment conditions. The basic entry requirements are a teacher qualification certificate issued by the national government and corresponding pre-service study preparation. At present, pre-service learning and training of mathematics teachers in primary and secondary schools are mainly achieved through pre-service education programs in universities.

When the People's Republic of China was first founded, responsibility of primary school teacher education was mainly through the establishment of technical secondary schools called Secondary Normal Schools. These schools recruited qualified junior high school graduates nationwide and trained them to become qualified primary school teachers. They implemented a three- or four-year schooling system. At that time, Secondary Normal Schools were the main sites for educating primary school teachers. Once these teachers have started their teaching careers in the classroom, they often partook in professional development programs to advance their degrees.

In the mid-1980s, a five-year college was established to educate primary school teachers and secondary school teachers. With the advancement and development of science and technology, as well as the society's demand for teachers, the pre-service education programs have transitioned from simply educating a large quantity of teachers to focus on the quality of teacher education. In 1999, China began to explore the training of undergraduate students in preparation for teaching in primary schools.

The education standards of elementary school teachers are also gradually improving. In the past, elementary teachers could have been trained at a Secondary Normal School, Normal College, or Normal University. Since 1999, there have been two main modes of elementary teacher training: Normal College and Normal University undergraduate. More recently, the level of elementary teacher education has risen again to encompass mostly teachers who have an undergraduate degree in education from a Normal University, a subject-specific undergraduate from a regular university, or a master's degree. The teacher training centers and Normal Universities have also greatly improved the quality of education that pre-service teachers receive.

At present, the primary mathematics teacher education program is part of the primary teacher education program at a Normal University. According to the classification by Ma (2018), there are currently three types of training models: comprehensive model, intermediate model, and subject model. The setting of pre-service courses for primary school mathematics teachers varies due to the different training models. The proportion of mathematics knowledge-related courses in the subject intermediate model is significantly higher than those of the comprehensive model. Pre-service teachers who are part of the comprehensive model program are provided with two to three courses in advanced mathematics, mathematical foundation, and elementary number theory, while those who are part of the subject-specific and intermediate model program are provided with multiple compulsory courses and a certain number of elective courses to help students in mathematical knowledge.

Courses in mathematical analysis, spatial analytic geometry, elementary number theory, non-Euclidean geometry, higher algebra, probability theory, projective geometry, calculus, real numbers and series, linear algebra and analytic geometry, probability statistics, a brief history of mathematics, mathematical thinking methods, mathematical culture, mathematical modeling, mathematical software, and mathematical experiments are relatively abundant, but there is no uniform requirement for the curriculum of mathematical knowledge for pre-service teachers. The teachers who teach mathematics knowledge courses in the teacher education program mainly come from the Department of Mathematics at the university and have profound and professional mathematics knowledge, or they come from a school of education with a background in mathematics.

The research shows that teachers and students generally find that the teachers from the Department of Mathematics are more successful in providing a strong mathematics program for the pre-service teachers (Xie, Ma, & Chen, 2018). These instructors are often able to establish a deeper relationship between advanced subject knowledge and subject knowledge that is taught in primary school. In addition, these instructors are able to form a knowledge network and establish vertical and horizontal connections among various types of teacher knowledge.

Generally speaking, the curriculum structure of primary mathematics teachers training at different institutions is largely similar. There is often not much difference in the proportion and content of the general curriculum and educational practice curriculum. The proportion of course types in different training modes differs mainly in the comprehensive model teacher training program as there is a significantly greater number of educational theory courses. In general, this part of the curriculum mainly includes general psychology, developmental psychology, and educational psychology, education introduction, teaching theory, curriculum theory, history of Chinese education, history of foreign education, education research methods, education statistics, education management, education philosophy, and principles of moral education. Using the mathematics module as an example, the "intermediate" subject curriculum is mainly composed of primary mathematics courses but also includes courses in science; thus, its proportion of subject-specific courses is higher than the other two models.

Some pre-service teacher education instructors believed that primary school mathematics teachers should have a complete and systematic subject knowledge system. The current primary school mathematics teachers also come from the School of Mathematics of a normal school most often teach in secondary schools. For Northeast Normal University's national funded teachers, the School of Mathematics has provided systematic and comprehensive courses in mathematics and education theory. They offered advanced mathematics courses such as higher algebra, complex function, and modern functions, which comprises almost 50% of their course load, while the ratio of teacher education courses was 16%. It shows that the school of mathematics aims to train the pre-service teachers with basic educational theory and methods, as well as with solid subject-based knowledge and competency. In addition, the teachers graduated from this school qualify to engage in subject research and teaching.

While subject-specific courses are important to increase subject-based knowledge and competency, pre-service programs need to ensure that educational theories, philosophy, and pedagogies are the core of these courses (Shi, 2007). In comparison with the primary education major teachers, the mathematics teachers educated by the School of Mathematics often have increased mathematical literacy and a solid mathematical theoretical foundation; often, they are more inclined to teach at middle schools.

All of the pre-service teachers were required to have at least eight weeks of school-based observation and two months practicum during their pre-service program. They have to pass the national mathematics teacher qualified certification examination before starting their career.

3 Pre-Service Mathematics Teacher Education in Canada

Canada is the only industrialized country where education is exclusively within the jurisdiction of the provinces. As there is no national department of education, the provincial government is responsible for the selection and education of teachers. In many provinces, that responsibility has been delegated to another organization. There are 10 provinces and three territories in Canada and they each have a department or ministry of education. They are responsible for the organization, design, delivery, and assessment of elementary, secondary, and tertiary levels. This responsibility includes the design and implementation of the provincial curriculum.

In many provinces, school districts were created to be responsible for the hiring of teachers and the delivery of the curriculum in their jurisdiction. In Ontario, there are 72 school districts. There are 38 public school districts (with four French school districts) and 38 public Catholic school districts (with eight French school districts and one Protestant school board). There are also seven school authorities responsible for children's treatment centers in Ontario. The district school boards are responsible to hire teachers and other staff to operate the school, to provide special programs, build schools, and to ensure that the provincial education law is followed.

In Ontario, the Ontario College of Teachers (OCT) sets the guidelines for certification and registration, and the universities are responsible for delivering the education program. In 2015, all teacher education programs were changed to be at least 16 months in length from a minimum of eight months in length. There are two types of educational programs for pre-service teachers: consecutive and concurrent (Gambhir, Broad, Evans, & Gaskell, 2008). The consecutive model is normally a two-year program with a prerequisite of a completed four-year undergraduate university degree. This route makes it possible for mature and experienced professionals to undertake a degree in teaching. It assumes that studies in many different subjects are an appropriate foundation for studies in education. The concurrent program model is designed so that students pursue studies in teaching at the same time as their academic studies. A common pattern is for teacher education to be integrated into the first or second year of a four- or five-year degree. The Faculty of Education works collaboratively with other undergraduate departments in developing and delivering the program.

In Ontario, there are different degrees available to students. The most common degree is the Bachelor of Education (B.Ed.) degree. This undergraduate degree is recognized by the Ontario College of Teachers as meeting the professional requirements and accreditation requirements to receive teaching certification in Ontario. There are also two graduate teacher education programs in Ontario, both at the Ontario Institute for Studies in Education at the University of Toronto. Both the Master of Teaching degree and the Master of Arts in Child Study degree provide graduates with a teaching certification in the province of Ontario and a master degree. These programs are 20 months in length and provide teacher education courses at the graduate level.

There are 16 universities certified in Ontario to provide teacher education programs. A university receives permission by the Ontario Ministry of Education to offer teaching certification after satisfying the accreditation requirements set by the Ontario College of Teachers. Every eight years, the Faculty of Education at each university must be recertified by the Ontario College of Teachers. While there are different teacher education models at these universities, the teacher candidates must have at least 80 practicum days in an Ontario school under the direction of an Ontario College of Education certified teacher. All teacher education programs require some form of teaching experience in schools. The experienced teacher in the classroom mentors the teacher candidate, observes their teaching, and provides feedback on the teaching practice of the teacher candidate.

The admission to teacher education programs is different for every university. To become qualified to teach mathematics at the middle (Grades 7 to 10), potential teacher education candidates need at least six half-course in mathematics at the university level. To teach mathematics at the senior school level (Grades 11 and 12), potential teacher candidates need to have at least twelve half-courses in mathematics in their undergraduate degree. However, the potential teachers do not need any courses in university mathematics to teach primary grades (K to Grade 6).

The teacher education models are affected by the curriculum, vision of the program, and areas and levels of professional expertise of the teacher educators (Van Nuland, 2011). The students take courses in foundations of education and methods. For potential elementary school teachers, the method courses include subjects such as mathematics, language, science, social sciences (history and geography), art, music, physical education, drama, and visual arts. All students take at least 36 hours of mathematics instruction in preparation to teach in elementary schools in Ontario. The students also take courses whose content is in psychology, special education, assessment, classroom management, English as a second language, and aboriginal education. In order to graduate with an Ontario College of Education teacher certificate, all teacher education candidates must successfully complete the Math Proficiency Test. The Ontario Government implemented this test in 2019-2020.

For teacher candidates planning to teach middle or senior school, they take additional courses to qualify them for teaching the more senior grades. For a specialist in middle school mathematics, the teacher candidate will take an additional course of 36 hours in mathematics teaching methods. This extra course is directed toward teaching and learning mathematics for Grades 7 to 10 students. For the teacher candidates who wish to teach senior grades in secondary schools (Grades 11 and 12), there is an additional course with 72 hours of mathematics instruction. They do not take the elementary school mathematics course of 36 hours.

There are a number of challenges to teaching people to be teachers of mathematics. In Ontario, there are very few requirements related to taking courses in mathematics in university. Some researchers and teacher educators have suggested that all teacher candidates be required to take at least one undergraduate course in mathematics (Kajander, Kotsopoulos, Martinovic, & Whiteley, 2013). Some Faculties of Education are introducing extra math courses and math tests to assist in increasing the content knowledge of teachers (Reid & Reid, 2017). Access to teacher education programs is challenging for some of the underrepresented groups in the multicultural province of Ontario.

These challenges have been tackled by a number of organizations and educational groups. The Canadian Deans of Education have developed a General Accord to advance Canadian Education. They have also created an Accord on Initial Teacher Education (ACDE, 2017) to articulate the goals, values, and principles to guide a national discussion on teacher education. They believe that there is an intellectual and practical component to teacher education and this should be situated in a university context. They identified three principles to guide teacher education in Canada. The three principles are:

- *Principle 1*: Teacher education programs prepare professional educators who effectively and skillfully foster learning;
- *Principle 2*: Teacher education programs prepare professional educators who engage in responsive and responsible collaboration; and
- *Principle 3*: Teacher education programs prepare professional educators who foster social responsibility.

These three principles are designed to assist teacher education program designers and implementers to better reflect the goals of a Canadian approach to teacher education where there is no federal organization.

4 Professional Development of In-Service Mathematics Teachers in Mainland China

Teacher training in China has gone through four stages from 1978 to the present: compensatory training, exploratory continuing education, universal continuing education, and national training (Chen & Wang, 2013). Primary and secondary school mathematics teachers need the training of teachers' colleges or universities and cannot do without the unique in-service teacher training opportunities. China attaches great importance to the development of in-service teacher training. Since 2008, the central government has increased investment in in-service teacher training at the compulsory education stage, and schools have actively explored and tried various professional development models based on actual conditions.

Various schools organize the education of new teachers differently. They provide support for teachers through on-the-job training, by pairing experience exemplary teacher with new teachers, school-based training, collective teaching, and research projects. Some scholars have divided the types of professional communities into their different dimensions of learning methods (theory or practice, online or offline) and regions: school-based teacher community and inter-school teacher community; teacher learning community and teacher practice community; virtual teacher community and non-virtual teacher community (Wang & Song, 2010).

Through the community platform established by the school, teachers at different levels can continuously reflect on their own methods, and content for learning and teaching (Quan, 2009). The establishment of professional communities has become an important form of promoting teacher professional development.

Another type of learning community revolves around outstanding teachers in a school, district or province. These outstanding teachers are often exemplary not only in their teaching but also in their understanding of students, research in education, and active professional development. As a leader for a learning community, these teachers act as role models in leading other teachers through their teaching methods and research practices in hopes of providing useful learning experiences for other teachers.

4.1 National Training Plan in China

In recent years, teachers have participated in "national training plan" carried out by various universities via the National Training Program website. For example, Northeast Normal University's seminar course, *Professional Development of Principals of Rural Schools*, organized by the Faculty of Education at Northeast Normal University, is part of the "National Training Program for Primary and Secondary School Teachers," which was fully implemented by the Ministry of Education and the Ministry of Finance in 2010. Its purpose is to improve the overall quality of primary and secondary school teachers, especially rural teachers. The "National Training Program" includes two items: "model training programs for teachers in primary and middle schools" and "training programs for backbone teachers in rural areas in the central and western regions." It is a model training program for primary and secondary school teachers in various provinces (autonomous regions, municipalities). Specific professional development opportunities are organized for different groups of educators such as principal training, backbone teacher training, and new teacher training. These multilevel professional development opportunities are rich in content and diverse in form (Table 1).

Chen and Wang (2013) found that most teachers believe that participation in training programs at normal universities or colleges and the training of provincial teachers' training colleges are better than other

Table 1 Types and content of in-service teacher training

Type of training	*Details*	*Details of the training*
Core teacher training	Principal training Exemplary teacher training Middle-level core teacher training	Teacher professional development Teaching reform Educational theory
Teacher training	New teacher training Young and middle-aged teacher training Old teacher training	Subject teaching, teacher teaching skills Student development and student management Class teacher work, teacher professional development

forms. They also found that 23.7% of teachers believe that the most effective support for their work is first and foremost, district's teaching activities. Other useful activities include focused study and research during school breaks, in school research, and field studies at different schools (Chen & Wang, 2013). These teachers are less convinced by the effectiveness of university-based professional development. Indeed, some teacher education activities, such as school-based training, school district teaching, and research, can more easily be integrated into a teacher's daily practice and work, and thus can fully meet the school requirements and teacher needs for professional development.

The professional development of a teacher is a process in which an individual teacher relies on professional organizations to acquire professional knowledge and skills through lifelong professional training throughout their entire career (Ye et al., 2001). Teachers' professional development is a successful strategy to promote the development of teacher education and improve teachers' social status (Zhong & Hu, 2005). Teachers' professional development standards are standards that guarantee the basic quality of teachers and a benchmark for engaging in education and teaching. Teachers as professionals must be trained on the basis of certain standards.

To this end, in order to deepen the reform of teacher education, standardize and guide teacher education courses and teaching, and cultivate high-quality teachers, the Ministry of Education issued the *Teacher Education Curriculum Standards* in 2011 (Government Portal of the People's Republic of China), followed by the *Professional Development Standards and Guidance for Primary and Secondary School Teachers* in

September of 2012. Teacher Education Curriculum Standards focuses on the professional development of teachers and specifies the goals and content of teacher education courses. It is the basis and guidance for teacher education courses created by various types of teacher education institutions at all levels to carry out teacher education and reform. The establishment of such teacher education standards can promote teacher education curriculum reform and improve teachers' professional quality and ability.

The development of the *Professional Development Standards and Guidance for Primary and Middle School Teachers* (Task Force, 2013) took three years. The guidelines are divided into two dimensions: professional foundation and professional practice. Teacher professional development standards were compiled according to nine subject areas, of which mathematics teacher professional development standards were divided into four elements. The professional foundation in this field includes: (1) sound personality and professional ethics; (2) professional knowledge in disciplines and education; (3) professional practice including the promotion of student learning and development; and (4) education and teaching research and professional development (Table 2).

The four areas are further subdivided into 20 standards: mastering knowledge about mathematics and adjacent subjects, students, mathematics courses, mathematics teaching, science and humanities, being able to design reasonable mathematical teaching programs, and implementing effective mathematics teaching activities and other standards. Mathematics teaching is a specialty, and teachers who are engaged in mathematics teaching are professionals. Their training and education must develop along the direction of specialization and professional standards.

Each standard in the guideline document is divided into three stages of "novice to proficient," "proficient to mature," and "mature to excellent" according to how the teacher develops. The requirements of each stage are different and correspond to detailed results indicators. For example, taking mathematics as an example (Table 3), the standard clearly specifies the requirements that mathematics teachers should meet at three different levels of subject knowledge. That is, novice teachers only need to master the knowledge of mathematics for the current semester they are teaching.

At the maturity stage, we should further understand the knowledge of adjacent semesters. Mathematics teachers are required to have a systematic grasp and also a profound and comprehensive knowledge of mathematics. That is, from the beginning of understanding to grasping to the formation

Table 2 Primary and secondary school teachers' professional development standards and guidelines: mathematics teacher framework

Professional foundation	Complete personality and professional ethics	Dedicated to my job and fulfilling my duties Caring for students and teaching Model for others, rigorous study Love life, physical and mental health
	Subject and educational teaching expertise	Subject knowledge Knowledge about students Knowledge about the course Knowledge about teaching and subject teaching Science and humanities
Professional practice	Promote students' learning and development	Create a good learning environment Design a sound teaching plan Implement effective teaching activities Develop good study habits and guide students to learn Carry out multiple learning evaluations Promote effective classroom management and infiltrate ideological and moral education and life skills education Implement active safety and health education
	Educational teaching research and professional development	Educational teaching reflection and action research Cooperation and experience sharing Lifelong learning and sustainable development

of a dynamic knowledge structures and problem-solving ability, a teacher's knowledge deepens layer by layer and requires progressive improvement.

In 2017, the Chinese Ministry of Education issued *The Training Curriculum Standards Guidance of Compulsory Education in Mathematics Teaching*. It is designed based on teachers' professional standards, mathematics curriculum standards, teacher education curriculum standards, and

Table 3 Primary and secondary school teachers' professional development standards and guidance indicators on the subject knowledge of mathematics teachers

Standard	*Outcome indicators*		
	From novice to proficient	*From proficient to mature*	*From proficiency to excellence*
5. Knowledge about the subject	5.1 Master the knowledge of mathematical disciplines specified in the mathematics curriculum standards for this semester 5.2 Understand the knowledge system of mathematics in this semester	5.1 Master the mathematics knowledge clearly specified in the mathematics curriculum standards for this semester. Understand the knowledge of adjacent semesters related to this semester 5.2 Grasp the knowledge system of mathematical disciplines and be able to describe the hierarchical structure of discipline knowledge 5.3 Understand the relationship between mathematics and other disciplines, as well as life practice and social development, and is able to consciously penetrate into mathematics teaching	5.1 Mathematics subject knowledge with dominance, profoundness, and applicability 5.2 In the rich subject practice experience, form a dynamically developing subject knowledge structure and the ability to solve practical problems 5.3 Have multidisciplinary knowledge and can be well integrated

national documents related to teacher training. This training curriculum standards clarifies training for specific topics such as numbers and algebra, figure and geometry, statistic and probability, synthesis, and practice. Its overall aim is to improve teachers' professional development.

5 Professional Development in Ontario

There are a number of ways that teachers in Ontario can continue to develop as teachers. There is a new teacher induction program, Additional Qualification courses, and graduate study programs. School districts are required to provide induction and mentoring programs for new teachers. These induction programs include formal school board orientations, individual school-level orientations, and professional development on safe schools, student success, classroom management, teaching students with special needs, and communicating with parents (Ontario College of Teachers, 2011). The continuum that begins in the faculty of education and continues through practice teaching and into a teaching position is enhanced by a teacher induction program (Glassford & Salinitri, 2007).

The new teachers are assigned an experienced mentor teacher to provide "coaching, information and demonstration of teaching methods" (Ontario College of Teachers, 2011, p. 14). New teachers have rated mentoring very high compared to induction programs (Crocker & Dibbon, 2008). This makes sense as mentoring builds a personal relationship with individuals while induction programs are usually about orientation.

The Ontario College of Teachers applies the same standards of practice to all teachers, regardless of their experience. Therefore, new teachers are assessed and evaluated on the same standards of practice as very experienced teachers. There are also occasional teachers in schools in Ontario, and the induction support for those teachers is not consistent. In other jurisdictions, there are different expectations for new and experienced teachers and so the induction programs will look very different.

In-service teachers may also take Additional Qualification courses through faculties of education. These courses are regulated and accredited by the Ontario College of Teachers. Teachers wishing to teach middle school mathematics but graduated with qualifications in Primary/Junior (Grades K to 6) may take a course called Additional Basic Qualifications Intermediate Mathematics to receive the qualifications to teach mathematics in Grades 7 to 10. Teachers, who wish to teach Senior Mathematics after getting the teaching certification, can take the Additional Basic Qualifications Senior Mathematics course to gain certification to teach Grades 11 and 12 mathematics courses. There are also courses available for teachers of Grades K to 6 called Three-Par Additional Qualification courses. The Mathematics, Primary and Junior course is taught at the Part

1, Part 2, and Specialist level online and in face-to-face delivery across Ontario.

Teachers may also enroll in graduate programs at local, national, and international universities. These programs provide Master of Education and Master of Arts degrees at the master's level. These programs are typically one to three years in length depending on the number of courses completed each year and the thesis component of the degree. There are also Doctor of Education (ED.D.) and Doctor of Philosophy (Ph.D.) degrees available in education at many universities. These degree programs provide professional and research experiences for students that will enhance the teachers' understanding of educational theory, practice, and research. These courses and degrees can also be used for prerequisites for taking Principal qualifications. Universities through the Additional Qualifications program teach these qualification courses.

6 Conclusion

In this chapter, we explore mathematics teacher education for both pre-service and in-service teachers. We identify the qualifications for potential teachers to enter programs, the courses that they take, and the professional development opportunities that they have after they graduate and become teachers. There are some similarities and many differences. It is clear that mathematics teachers in China have more mathematics experience and a more robust professional development plan than teachers in Ontario.

References

Association of Canadian Deans of Education. (2017). *Accord on initial teacher education*. Retrieved from: http://www.csse-scee.ca/docs/acde/acde_teacher educationaccord_en.pdf.

Chen, X., & Wang, Z. (2013). Survey on teacher training in compulsory education: Status problems and suggestions. *Open Education Research, 4,* 11.

Crocker, R., & Dibbon, D. (2008). *Teacher education in Canada*. Kelowna, BC: Society for the Advancement of Excellence in Education.

Gambhir, M., Broad, K., Evans, M., & Gaskell, J. (2008). *Characterizing initial teacher education in Canada: Themes and issues*. Prepared for the International Alliance of Leading Education Institutes.

Glassford, L., & Salinitri, G. (2007). Designing a successful new teacher induction programme: An assessment of the Ontario experience, 2003–2006. *Canadian Journal of Educational Administration and Policy, 60*, 65–74.

Government Portal of the People's Republic of China. http://www.moe.gov.cn/.

Kajander, A., Kotsopoulos, D., Martinovic, D., & Whiteley, W. (2013). *Mathematics pre-service teacher education in Ontario.* Consultation brief regarding the extended pre-service program. Toronto, ON, Canada: Fields Institute.

Ma, Y. (2018). Viewing the training of excellent primary school teachers from the positioning of primary education majors. *Journal of Northeast Normal University (Philosophy and Social Sciences), 4*, 143.

Ministry of Education. (2011). *Teacher education curriculum standards.* Retrieved from: http://www.moe.gov.cn/.

Ministry of Education. (2012). *Professional development standards and guidance for primary and secondary school teachers.* Retrieved from: http://www.moe.gov.cn/.

Ontario College of Teachers. (2011). *Ontario college of teachers transitions to teaching 2010: Early-career teachers in Ontario schools.*

Quan, L. (2009). Professional growth of teachers in the studio environment of famous teachers: A perspective of professional community. *Teacher Education, 13*, 31.

Reid, M., & Reid, S. (2017). Learning to be a math teacher: What knowledge is essential? *International Electronic Journal of Elementary Education, 9*(4), 851–872.

Shi, N. (2007). Some thoughts on mathematics curriculum standard. *Bulletin of Mathematics, 5*, 1.

Task Force on Professional Development Standards and Guidance for Primary and Middle School Teachers. (2013). *Professional development standards and guidance for primary and middle school teachers.* Beijing: Normal University Press, 213–225.

Van Nuland, S. (2011). *Journal of Education for Teaching, 37*(4), 409–421.

Wang, Y., & Song, T. (2010). A probe into the types of teachers' community in the new era. *Journal of Xinzhou Normal University, 2*, 130–132.

Xie, S., Ma, Y., & Chen, Y. (2018). Elementary mathematics teacher preparation in China. In Y. Li & R Huang (Eds.), *How Chinese acquire and improve mathematics knowledge for teaching* (pp 85–108). Leiden: Brill Sense.

Ye, L., et al. (2001). *New exploration of teacher role and teacher development.* Beijing: Education Science Press.

Zhong, O., & Hu, H. (2005). Construction of teacher education curriculum standards in China. *Global Education Outlook, 1*, 37.

CHAPTER 3

Mathematics Curriculum in Ontario, Canada and Mainland China

Sijia Cynthia Zhu and Shu Xie

1 Overview of Mathematics Curricula

In this chapter, we discuss the curriculum that guides mathematics education in Mainland China and Ontario, Canada. In Mainland China, there is a national curriculum standard while in Canada, the curriculum is the responsibility of the individual provinces and territories. In China, the compulsory education mathematics curriculum is divided into three phases covering grades 1–9 and was most recently updated in 2011. In Ontario, the current mathematics curriculum documents were most recently updated in 2005 and covers grades 1–8. While the curricula of the two regions have some shared objectives and expectations, there are also numerous differences in organization, philosophy, and usage.

S. C. Zhu (✉)
Ontario Institute for Studies in Education, University of Toronto, Toronto, ON, Canada
e-mail: cyn.zhu@mail.utoronto.ca

S. Xie
Faculty of Education, Northeast Normal University, Changchun, China
e-mail: 34474878@163.com

S. C. Zhu et al. (eds.), *Reciprocal Learning for Cross-Cultural Mathematics Education*, Intercultural Reciprocal Learning in Chinese and Western Education,
https://doi.org/10.1007/978-3-030-56838-2_3

2 Mainland China: Mathematics Curriculum for Compulsory Education

2.1 *Mathematics Course Objectives*

The *Full-time Compulsory Education Mathematics Curriculum Standard (2011)*, herein after referred to as the *Curriculum Standard*, had strict requirements on the overall objectives of the mathematics curriculum and had adopted a combination of general and specific methods. The general goals of the document included acquiring important mathematical knowledge, such as mathematical facts, experience in mathematical activities, basic mathematical thinking methods, necessary application skills, increasing awareness of applied mathematics, enhancing understanding of mathematics, and increasing confidence in learning mathematics. The concrete goals included knowledge and skills goals, mathematical thinking goals, problem-solving goals, and emotional attitude goals. These four concrete goals are a closely related organic whole and are realized in a variety of mathematical activities.

2.2 *The Function of Curriculum*

Yang (2010) demonstrated that the mathematics curriculum in compulsory education is a basic curriculum for cultivating citizens' qualities, and it is basic, popular, and developmental. Mathematics courses can enable students to master the necessary basic knowledge and basic skills, cultivate students' abstract thinking and reasoning ability, and to cultivate students' innovative awareness and practical ability, as well as promoting students' emotional development, attitudes and values. The mathematics curriculum of compulsory education can lay an important foundation for students' future life, work, and study.

2.3 *General Characteristics of the Mathematics Curriculum*

The mathematics curriculum should be dedicated to the realization of the training goals of compulsory education. It should be geared to all students and meet the needs of the students' individual development, so that everyone can get a good mathematics education, and each student can get an appropriate program in mathematics. The content of the mathematics curriculum should reflect the needs of society and the characteristics of mathematics, and it should conform to the cognitive laws of students. It should include not only the results of mathematics, but also the formation process of mathematical results and the implication of mathematical thinking methods.

The teaching activities should provide the students with a process of active participation, interaction, and common development. The main purpose of assessment is to fully understand the process, produce results of students' mathematics learning, motivate students to learn, and to improve teaching. An assessment system with multiple objectives and multiple methods should be established. Lastly, the design and implementation of mathematics courses should use modern information technology in a reasonable manner according to the actual situation. Teachers should pay attention to the integration of information technology and course content, and to actual student achievement results.

2.4 Structure of the Mathematics Curriculum

The design of the mathematics curriculum at the compulsory education stage should fully consider the characteristics of students' mathematics learning, which should be consistent with the students' cognitive laws and psychological characteristics, thus conducive to stimulating students' learning interest and triggering students' mathematical thinking. The curriculum considers the specific characteristics of mathematics learning and balances content knowledge with skills and processes. The design of the curriculum emphasizes students' prior experiences and uses examples, derived from real life, to demonstrate abstract mathematics concepts. Through this process, students are able to create mathematical models and gain a deeper understanding of problem solving (Ministry of Education of the People's Republic of China, 2012). According to the integrity of the mathematics curriculum and the psychological and physical characteristics of the students, the students' mathematics learning is divided into three periods: grades 1–3, grades 4–6, and grades 7–9. Its goals were mainly elaborated from four aspects: knowledge and skills, mathematical thinking, problem solving, and emotional attitudes.

The course content mainly includes four strands: numbers and algebra, graphics and geometry, statistics and probability, and synthesis and practice. The main content of N*umbers and Algebra* is the understanding of numbers, representation of numbers, size of numbers, operation of numbers, estimation of numbers, using letters to represent the numbers, algebraic expressions and their operations, equations, systems of equations, inequalities, and functions. The main focus of *Graphics and Geometry* is to understand the basic graphics in space and plane, nature, classification and measurement of graphics, as well as the translation, rotation, axisymmetric, similarity, and projection of graphics. Additionally, *Graphics and Geometry* covers topics such as the proof of the basic properties of planar graphics, and the use of coordinates to describe graphics' location and movement. The focus of the *Statistics and Probability* strand is to teach students to collect, organize, and describe data. This includes simple sampling, organizing the data collected, and creating graphs to represent the data. Students should also learn to analyze data by calculating mean, median, mode, and variance. Lastly, through the statistics and probability strand, students should learn to use data to make prediction and estimate probability.

The purpose of the *Synthesis and Practice* strand is to teach students to comprehensively apply relevant knowledge and methods to solve practical problems, to cultivate students' problem awareness, application and innovation awareness, accumulate students' activity experience, and to improve students' ability to solve real problems. *Synthesis and Practice* is a type of learning activity that uses mathematics problems to focus on students' participation. In the learning activities, students will comprehensively use knowledge and methods learned through the other three strands—number and algebra, graphics and geometry, and statistics and probability—to solve problems.

In the mathematics curriculum, attention is given to the development of students' sense of numbers, symbol consciousness, concept space, geometric intuitiveness, ability to analyze data, computing ability, reasoning ability, and mathematics model thinking. In order to meet the needs of students, the mathematics curriculum pays special attention to the development of students' sense of application and innovation.

2.5 *The Goals of Mathematics Curriculum*

The goals of the compulsory mathematics education suggest that students should gain the ability to:

- Acquire basic knowledge, basic skills, basic thoughts, and basic activity experience of mathematics necessary to adapt to social life and further development.

- Experience the connection between mathematics knowledge, mathematics and other disciplines, mathematics and life, and to use mathematical thinking to think. This should enhance a student's ability to find and ask questions, analyze and solve problems.
- Understand the value of mathematics, as well as increase their interest in learning mathematics, enhance the confidence to learn mathematics well, develop good learning habits, and have a preliminary sense of innovation and scientific attitude.

The goals are placed under four categories: knowledge and skills, mathematical thinking, problem solving, and emotional attitude (Table 1).

For each semester, there are specific syllabus goals. For example, in the *knowledge and skills* category in grades 1–3, students should experience the process of abstracting numbers from real life, understand the meaning of numbers within 10,000, and get a preliminary understanding of fractions and decimals. Students should understand the meaning of common quantity, and to understand the meaning of the four operations, as well as having the ability to master the necessary computing skills, and be able to accurately calculate. Given a specific situation, students should be able to choose the appropriate unit for simple estimation.

2.6 *Assessment*

According to China's mathematics curriculum standard, the main purpose of assessment is to comprehensively understand the process and results of students' mathematics learning, to motivate students to learn, and to improve teaching. Assessments should be based on the course objectives and course content, reflect the basic concepts of the mathematics curriculum, and comprehensively assess students' performance in knowledge and skills, mathematical thinking, problem solving, and emotional attitudes.

Assessment should focus on students' learning results and on students' development and changes in the learning process. Diversified assessment methods should be adopted, the assessment results should be properly presented and reasonably used, and the incentive effect of assessment should protect the self-esteem and self-confidence of students. Through the information obtained through evaluation, the level of students' mathematics learning and existing problems can be understood. Assessment

Table 1 Four strands of mathematics content in the China mathematics curriculum

Knowledge and Skills	Experience the processes of abstraction, operation and modeling that belong to algebra, master the basic knowledge and basic skills of numbers and algebra Go through the abstraction, classification, nature discussion, movement, position determination and other processes of graphics, master the basic knowledge and basic skills of graphics and collections Experience the process of collecting and processing data in actual problems, using data to analyze problems, and obtaining information, and master the basic knowledge and basic skills of statistics and probability Participate in comprehensive practical activities and accumulate experience in mathematical activities that comprehensively use mathematical knowledge, skills and methods to solve simple problems
Mathematical thinking	Establish sense of numbers, symbol consciousness and space, initially form geometric intuition and computing ability, develop image thinking and abstract thinking Experience the significance of statistical methods, develop data analysis concepts, and experience random phenomena In participating in mathematical activities such as observation, experiment, conjecture, proof, and comprehensive practice, develop the ability of reasonable reasoning and deductive reasoning, and express ideas clearly Learn to think independently and experience the basic ideas and ways of thinking in mathematics
Problem solving	Initially learn to find problems and raise problems from the perspective of mathematics, comprehensively apply mathematical knowledge to solve simple practical problems, enhance application awareness, and improve practical ability Get some basic methods for analyzing and solving problems, experience the diversity of problem-solving methods, and develop innovative awareness Learn to cooperate and communicate with others Initially form the consciousness of evaluation and reflection

Emotional attitude	Actively participate in mathematical activities The curiosity about mathematics In the process of mathematics learning, to experience the joy of success, exercise the willing to overcome difficulties, and to build self-confidence Experience the characteristics of mathematics and understand the value of mathematics Develop learning habits of serious diligence, independent thinking, cooperative communication, reflection, and questioning Form a scientific attitude of upholding truth, correcting mistakes, and being rigorous and realistic

should help with teachers' summarizing and self-reflection, adjusting and improving the teaching content, and teaching process.

3 Ontario's Elementary Grade 1–8 Mathematics Curriculum

The Ontario curriculum was originally rooted in the colonial perspectives of Great Britain and France (Laidlaw, Davis, & Sumara, 2001). However, the Ontario curriculum has also always been influenced by American education ideologies. The influences of Americans educationalists, John Dewey and Howard Gardner, are found in the Ontario mathematics curricula. As an educationalist, Dewey (1966) believed that schooling and learning should be an interactive and social endeavor with an end goal for "social progress and reform" (p. 44).

Expanding upon Dewey's early theories, Jean Piaget conducted research specifically on how children function as constructivist learners and came to three main conclusions. First of all, Piaget (1954) suggests that children's minds are like structures and as they learn and grow building blocks are added to these structures thus enabling them to make sense of their outside world. Secondly, these building blocks, otherwise known as schemata, function as a bank of knowledge that children can use to deal with new situations through a process Piaget (1954) defined as assimilation.

Ontario math education and math teaching can be interpreted as an active process, which has its roots in constructivism. This theory of learning states that schooling should be a student-centered endeavor, which focuses on allowing the student to build and develop their own knowledge (Marlowe & Page, 1998). In order for students to cultivate new understandings, they must be given opportunities to explore and investigate the new knowledge that they are exposed to. Furthermore, according to constructivist theory, teachers should take on the role of a facilitator or guide in the learning process, providing students with strategies to connect their prior knowledge with new learning in hopes of gaining a deeper understanding (Cobb, Yackel, & Wood, 1992).

While the Canadian curriculum has been "uniquely positioned to take advantage of theoretical tools from the U.S" (Laidlaw et al., 2001, p. 144), it also has its own distinctive features. When trying to make a list of what is Canadian, Laidlaw et al. (2001) suggest that "an essential quality of Canada is the lack of essential qualities" (p. 145). In fact, the way that Canadian curriculum has developed recently shows that it deliberately values diversity and looks to draw upon the nation's people, geography, and multiculturalism (Laidlaw et al., 2001; Ontario Ministry of Education, 2005).

In addition, Canadian curriculum takes on a more ecological approach through its emphasis on relationships between society, the environment, and the people (Laidlaw et al., 2001; Ontario Ministry of Education, 2005). Examples of this can be seen through the elementary mathematics curriculum as it asks students to engage with their environment through observation and data.

3.1 Ontario's Current Curriculum

The Ontario Ministry of Education released their latest Elementary Mathematics Curriculum in 2005. This 138-page curriculum document caters to students from grades 1–8, emphasizing the constructivist viewpoints that "all students can learn mathematics and deserve the opportunity to do so through active learning that allows them to experiment and compare the predictions to the results" (p. 96) and through teachers using "culturally diverse materials" (p. 28). Furthermore, it recognizes that "all students do not necessarily learn mathematics in the same way using the same resources and within the same time frames" (p. 3).

The curriculum also emphasizes that experiences and applications of facts, skills, and procedures are necessary if students are to "learn mathematics in a way that will serve them well throughout their daily lives" (Ontario Ministry of Education, 2005, p. 3). One of the most prominent features in recent curricula is the focus on inquiry and problem-solving-based education. The Ontario grades 1–8 mathematics curriculum has sections titled "Literacy and Inquiry/Research Skills" and "application" which capitalize on the concepts of experiential learning and dialectical constructivism (Dewey, 1966; Moreno-Armella & Waldegg, 1993; Moshman, 1982; Ontario Ministry of Education, 2005; Vygotsky, 1978).

The curriculum document is in two main sections: the first 30 pages focus on pedagogical aspects of mathematics teaching and learning while the later 80 or so pages outline the expectations divided by grade and then strand. For each grade, students' learning is organized into five strands, which are Number Sense and Numeration, Measurement, Geometry and

Spatial Sense, Patterning and Algebra, and Data Management and Probability. All five strands are taught each year according to a set of overall and specific expectations that build upon the previous years.

3.2 Curriculum and Pedagogy

According to the Ontario Ministry of Education (2005), seven mathematical processes form the foundation for student learning in elementary mathematics and are meant to be taught by teachers and applied by students through all grades and strands. These seven processes include problem solving, reasoning and proving, reflecting, selecting tools and computational strategies, connecting, representing, and communicating. Throughout the school year, students should actively engage in these processes as they learn are introduced to new mathematical knowledge specific to their grade. The curriculum emphasizes that these processes are interconnected and cannot be separated from the learning of mathematics content. By engaging in each process, students will gain a deeper understanding of how their newly acquired knowledge can be applied.

The curriculum breaks down the seven processes in detail, giving educators explanations as to why each process is important to the learning of mathematics. For each process, specific examples are also provided as a guideline for teachers. An example of this is found in the section explaining the process of representing:

> a student in the primary grades should know how to represent four groups of two by means of repeated addition, counting by 2's, or using an array of objects. The array representation can help students begin to understand the commutative property (e.g., $2 \times 4 = 4 \times 2$), a concept that can help them simplify their computations. (p. 17)

While all seven processes are important for the learning of mathematics, the curriculum document emphasizes that the process of problem solving is central. According to the Ontario Ministry of Education (2005), learning to solve problems and learning through problem solving allow students to connect with mathematical ideas and develop conceptional understanding. Thus, problem solving should be the foundation of any mathematics program.

The curriculum guides educators by recommending the use of George Polya's (1945) four-step model for problem solving: understand the problem, plan, execute, check and reflect. At each step, the curriculum outlines actions such as "reread and restate the problem" (p. 13) for making a plan, and "do the necessary calculations" for execute. In an effort to emphasize mathematics communication, the curriculum document also makes recommendation for different ways students can communicate their thinking at each step. While problem solving is crucial, the Ontario Ministry of Education (2005) also emphasizes that not all mathematics instruction can take place through problem solving as some knowledge such as mathematics conventions, symbols, and terms must be taught explicitly.

3.3 Assessment Within the Curriculum

Assessment of K to 8 students in Ontario is guided by two main documents: *Growing Success: Assessment, Evaluation, and Reporting in Ontario Schools* (Ontario Ministry of Education, 2010) and the Mathematics curriculum (Ontario Ministry of Education, 2005). The documents outline a two-dimensional assessment plan that categorizes student learning into four main areas: Knowledge and understanding, thinking, communication, and application. Students are assessed in these four areas on a scale with 4 levels. Level 1 achievement means that students have limited understanding of the knowledge and skill, level 2 represents some understanding, level 3 represents the provincial standard of having considerable understanding, and level 4 means students have a thorough understanding of the knowledge. Within each level, teachers will further evaluate their students as a −1, 1, +1, and so on for each level. These levels are then converted into letter grades for grades 1–6 and percentages for grades 7–12.

The "*Growing success*" document outlines three of ways of assessing student learning: assessment for learning, assessment as learning, and assessment of learning. In mathematics, assessment *for learning* ranges from diagnostic evaluations to formative activities where students receive descriptive feedback and coaching with the goal of improvement. For example, a student may receive a short quiz before the start of a unit so that both the teacher and student can better identify gaps of knowledge, strengths, and weaknesses. Assessment *as learning*, a process of formative assessment, focuses on fostering student's development of new mathematical knowledge through teacher presentation and modeling, as well as opportunities for student practice and structured opportunities for self-assessment.

The Ontario Ministry of Education (2010) emphasizes that "the primary purpose of assessment and evaluation is to improve student learning" (p. 6) and not as a conclusion to students learning and thus should mostly fall under assessment for and as learning. Assessment of learning in mathematics usually occurs toward the end of a unit of learning and may take the form of a test or project. In the process of assessment, it is crucial that all students, teachers, and parents are aware of the learning goals and what successful achievement of the learning goals looks like.

3.4 Considerations for Planning

While assessment and mathematical processes are the core of mathematics teaching and learning, there are other important areas that teachers must consider when designing their program. The Ontario Ministry of Education (2005) begins this section of the curriculum by emphasizing that, in a mathematics classroom, students will exhibit diverse way of learning. Therefore, it is important that teachers provide students with a variety of learning modes including but not limited to independent learning, teacher-directed learning, cooperative learning, group learning, and hands on experimentation. By providing a variety of learning modes, teachers will also help foster students' communication skills and ability to interpret the mathematics in their daily lives, which will promote a more positive attitude toward mathematics.

In addition to using a variety of instructional approaches when teaching, the Ontario Ministry of Education (2005) also provides educators with guidance on two other areas of pedagogy: (1) connecting mathematics with other areas of learning and (2) teaching mathematics to a diverse student population. The Ontario Ministry of Education (2005) encourages teachers to provide students with cross-curricular and integrated learning opportunities to students that incorporate related content and skills learned in other subjects. Cross-curricular lessons integrate content and processes shared between different subject areas while the curriculum defines integrated learning specifically as learning opportunities that allow students to work toward meeting expectations from two or more subjects. Learning opportunities can be both cross-curricular and integrated.

The other category of pedagogical advice focuses on addressing the diversity within Ontario's mathematics classroom. The curriculum emphasizes that all students should be provided with unbiased and equal opportunity to reach their full potential and that teachers should have high expectations for all students. In order to do this, it is important that the learning activities and resources used reflect the diversity of the student body whether that be in ability, culture, race, or experience. For students with exceptionalities, teachers must determine whether the student needs accommodations to meet the grade expectations or modification to the curriculum and grade level for them to have to successful learning experience.

According to *The Individual Education Plan (IEP): A Resource Guide* (Ontario Ministry of Education, 2004), a manual for designing education plans for students with exceptionalities, there are three types of accommodations, which allow students to participate in regular grade-level curriculum and demonstrate their learning independently. Mathematics teachers can make instructional accommodations in terms of methods of presentation, environment accommodations that best suit a specific student learning needs, or assessment accommodations. None of these changes should alter the curriculum expectations required of the student. If the students are unable to successfully learn the mathematics content for their grade level, modification can be made to a student's mathematics program by creating a program from the expectations outlined in other grades. These accommodations and modification may also apply to students who are English language learners. The new mathematics program created in collaboration between the teacher, student, and parents is considered a student's individual education plan.

3.5 Curriculum Expectations

In the second section of the Ontario Mathematics curriculum document, expectations are listed by grade. For each grade, the curriculum begins with a summary of each strand and then reiterating the mathematical processes. From grade to grade, these processes are similar as the goal is to have students apply them to new mathematical knowledge. The main section of each grade's curriculum is divided into five strands: Number Sense and Numeration, Measurement, Geometry and Spatial Sense, Patterning and Algebra, and Data Management and Probability.

For each strand, the curriculum outlines two to four overall expectations that students should gain mastery of by the end of the grade. It then breaks down these expectations into specific expectations. For example, in grade 3 measurement, the overall expectations "estimate, measure, and record length, perimeter, area, mass capacity, time, and temperature using standard units" (p. 57) are supported by a series of specific expectations such as "draw items using a ruler, given specific lengths in centimeters (sample problem: draw a pencil that is 5 cm long)" (p. 57) and "choose benchmarks for a kilogram and a litre to help them perform measurement tasks" (p. 58). Figure 1 shows an example of the breakdowns of grade 5 mathematics.

The five strands are considered to be the five major areas of knowledge and skills in the elementary mathematics curriculum. The goals

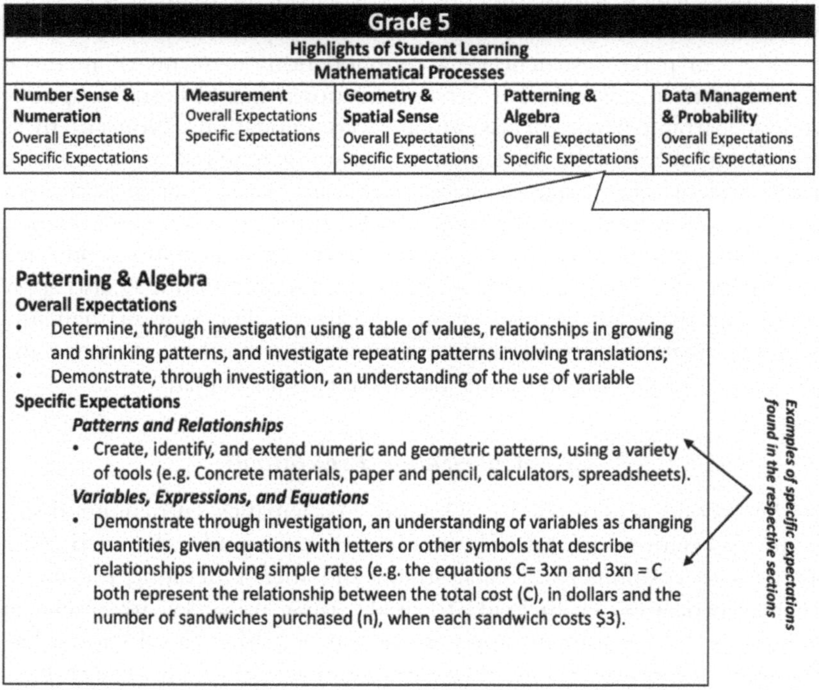

Fig. 1 Examples from the grade 5 mathematics curriculum in Ontario

for student learning through each strand, summarized from the Ontario curriculum, are as follows:

- *Number Sense and Numeration*: In this strand, students should develop a general understanding of numbers, their various representations, and relationships with each other. Further, students should gain a solid understanding of the four operations and learn to compute fluently using a variety of tools and strategies.
- *Measurement:* Student learning in this strand focuses on the measurable attributes of objects and the units and processes involved in measurement.
- *Geometry and Spatial Sense*: The goal for this strand is to develop students' intuitive awareness of their surroundings and the objects in them. Students should learn how to use geometric principles to represent and describe objects, and their interrelationship in space. This strand also covers topics such as location and movement.
- *Patterning and Algebra*: In this strand, students develop their knowledge of two core themes in mathematics: different ways of using numbers to represent equal quantities and patterns in shapes, designs, movement, as well as sets of numbers. When learning algebra, students will be introduced to variables and techniques for solving equations.
- *Data Management and Probability*: The goal for this strand is to equip students with the ability to gather, organize, display, and analyze data in different ways. Students should also experiment with probability and use probability models to simulate situations.

While these five strands are divided into individual sections for each grade, they are not meant to be taught in isolation. The curriculum recommends that, whenever possible, teachers should integrate the strands and apply them real-life situations. Furthermore, students should be given multiple opportunities throughout the school year to learn and demonstrate their understanding of a given concept. Examples of different activities that teachers can use in their teaching are provided for some specific expectations. Many of these examples integrate multiple subjects such as art, technology, and science.

3.6 *Controversy Over Ontario's Curriculum*

In recent years, Ontario's mathematics education has been under scrutiny as large-scale provincial test scores show that approximately 50% of grade 6 students and 38% of grade 3 students are not meeting the provincial standard of achieving level 3 in mathematics (EQAO, 2017). As this trend coincides with Ontario's mathematics curriculum, some argue that it is failing to build the strong foundational skills that students need to be successful (Alphonso, 2017). However, there are other researchers and teachers that question the effectiveness of standardized testing in determining students' achievement as these tests are similar to textbooks in their lack of attention to diversity of race, gender, ability, and class (Alphonso, 2017). Furthermore, Alphonso (2017) believes that there is an incongruity in the constructive, exploratory, inquiry-based ways students are taught in the classroom versus the strict individualized, pencil and paper standardized test.

Regardless of why students are scoring below average on the large-scale provincial tests, Ontario is implementing a Renewed Math Strategy that started in 2016 that will increase support for math teachers, opportunities for educators to deepen their knowledge, time for math instruction, and access to resources for parents (Zegarac, 2016). This renewed math strategy has come with additional documents to support teachers and parents. Furthermore, Ontario has been working on a new curriculum that is planned for release for fall 2020 that focuses more on foundational skills.

4 Conclusion

In this chapter, we have described the mathematics curriculum for mainland China and the province of Ontario. The Chinese curriculum has four strands of topics through grades one to nine, while the Ontario curriculum has five strands through grades one to eight. All of the Ontario strands are related to mathematics topics while one of the four Chinese strands is focused on synthesis and practice. This illustrates the importance of bringing the topics together and of mathematics practice that does not appear in the Ontario mathematics curriculum.

The Chinese curriculum standards are focused more on assessment of learning as a tool to understand student learning and for teachers' instructional improvement. On the other hand, the Ontario curriculum encourages investigation and inquiry and expects teachers to use a variety of assessment practices, including those for assessment *of*, *for*, and *as* learning. The variety of assessment in Ontario includes projects, presentations, quizzes, tests, performance tasks, and assignments.

Both educational jurisdictions recognize the importance of curriculum renewal as the current documents are becoming outdates. The two educational systems are preparing to release new curriculum documents that will provide teachers with teaching strategies and mathematics topics that will guide teaching practices into the future.

References

Alphonso, C. (August 30, 2017). Half of Grade 6 students fail to meet Ontario math standards. *The Globe and Mail*. Retrieved from https://www.theglobeandmail.com/news/national/half-of-grade-6-students-fail-to-meet-ontario-math-standards/article36124763/.

Cobb, P., Yackel, E., & Wood, T. (1992). A constructivist alternative to the representational view of mind in mathematics. *Journal for Research in Mathematics Education, 23*(1), 2–33.

Dewey, J. (1966). My Pedagogic Creed. In F. W. Garforth (Ed.), *Selected educational writings*. London, UK: Heinemann.

Education Quality and Accountability Office. (EQAO). (2017). *Highlights of the provincial achievement results, 2016–2017.* Retrieved from http://www.eqao.com/en/assessments/results/communication-docs/provincial-report-highlights-elementary-2017.pdf.

Laidlaw, L., Davis, B., & Sumara, D. (2001). Canadian identity and curriculum theory: An ecological, postmodern perspective. *Canadian Journal of Education, 26*(2), 144–163.

Marlow, B. A., & Page, M. L. (1998). *Creating and sustaining the constructivist classroom*. Thousand Oaks, CA: Corwin Press.

Ministry of Education of the People's Republic of China. (2012). *Mathematics curriculum standards for compulsory education (2011 Edition: 4–5)*. Beijing: Beijing Normal University Press.

Moreno-Armella, L., & Waldegg, G. (1993). Constructivism and mathematical education. *International Journal of Mathematical Education in Science and Technology, 24*(5), 653–661.

Moshman, D. (1982). Exogenous, endogenous and dialectical constructivism. *Developmental Review, 6*, 371–384.

Ontario Ministry of Education. (2004). *The Individual Education Plan (IEP): A Resource Guide.* http://www.edu.gov.on.ca/eng/general/elemsec/speced/guide/resource/iepresguid.pdf.

Ontario Ministry of Education. (2005). *The Ontario curriculum, grades 1–8: mathematics*. Toronto, ON: Queen's Printer for Ontario.
Ontario Ministry of Education. (2010). *Growing success, assessment, evaluation and reporting in Ontario schools: First edition, covering grades 1 to 12*. http://www.edu.gov.on.ca/eng/policyfunding/growSuccess.pdf.
Piaget, J. (1954). *The construction of reality in a child*. New York, NY: Basic Books.
Polya, G. (1945). *How to solve it*. Princeton, NJ: Princeton University Press.
Vygotsky, L. (1978). Interactions between learning and development. In *Mind and society*. Cambridge, MA: Harvard University Press.
Yang, Q. (2010). *Primary mathematics curriculum and teaching* (pp. 47–49). Beijing: Chinese People University Press.
Zegarac, G. (2016). *Ontario's renewed mathematics strategy*. Toronto, ON.

CHAPTER 4

Mathematics Resources in Ontario, Canada and Mainland China

Shu Xie, Sijia Cynthia Zhu, and Jiaxin Huang

1 Introduction

While curriculum documents are often seen as the main guide for mathematics education in both Canada and China, there are a large variety of resources that support educators in delivering that curriculum to students. In China, the main sources for curriculum-based teaching activities are textbooks and their companion workbooks. In Ontario, Canada, there is a larger variation in what is used by educators to deliver the curriculum.

S. Xie · J. Huang
Northeast Normal University, Changchun, China
e-mail: 34474878@163.com

J. Huang
e-mail: xin_694417040@qq.com

S. C. Zhu (✉)
Ontario Institute for Studies in Education, University of Toronto, Toronto, ON, Canada
e-mail: cyn.zhu@mail.utoronto.ca

S. C. Zhu et al. (eds.), *Reciprocal Learning for Cross-Cultural Mathematics Education*, Intercultural Reciprocal Learning in Chinese and Western Education,
https://doi.org/10.1007/978-3-030-56838-2_4

This chapter will showcase the various resources used in both Mainland China and Ontario, Canada.

2 Mathematics Textbooks in Mainland China

In the past, there were many mathematics textbooks developed by a variety of publishers. In recent years, the nation has been attempting to unify textbooks with the Ministry of Education as the only publisher. The Chinese Ministry of Education clearly states in the *Comprehensively Deepening the Curriculum Reform and Implementing the Fundamental Tasks of Morality Education* (Ontario Ministry of Education, 2014) that a unified national reform of three subjects' curricula, including morality and the rule of law, Chinese, and history, is required. In July 2017, the National Textbooks Committee was established to guide and coordinate the nation's textbook work, which indicated that the construction of textbooks in China had entered a new stage. However, currently there are many primary school mathematics textbooks, supporting the premise of one standard and multiple textbooks.

Based on the Chinese Mathematics Curriculum Standards, different versions of textbooks contain different writing styles and characteristics. At present, there are six versions of mathematics textbooks available in China. According to Ma (2012), the textbook published by People's Education Publishing House is the most common primary school mathematics textbook. This publishing house put great emphasis on science. The books are logically written and rigorously researched with a focus on fundamental skills.

Beijing Normal University Press' version of the mathematics textbook focuses on students' life experience and the connection between mathematics and life (Ma, 2012). Textbooks published by Jiangsu Education Press conform to the students' cognitive laws, highlighting the patterns within math pedagogy and the application of basic teaching methods. The examples, questions and problems chosen by the textbook, and the choice of content materials fully reflect the core values of socialism.

A Qingdao mathematics textbook version published by Education Press provides appropriate support for effective teaching in a context-based and Socratic methods. These teaching methods supported by the Qingdao publishing books help stimulate deep thinking in mathematics. Western Normal Edition textbooks published by Southwest Normal University Press are regionally developed books that focus on the needs of specific ethnic minorities. Lastly, Jijiao version textbooks published by Hebei Education Press include cartoon learning companions that follow students through the grades and help stimulate students' interest in learning. These textbooks also include many synthesis and practice activities for a variety of learning units.

This large variety of textbooks is born of China's curriculum reform in the early 2000s where the nation was no longer ascribing to a "one size fits all" form of education, but one that took into account the differences between the various regions and schools within the nation. Currently, different schools and regions can choose the textbook version suitable for their locale. This also allows teachers more flexibility in terms of how they design their mathematics teaching. Before selecting and using a specific resource in their teaching, teachers must first understand the theories, content, and pedagogies described by the textbook. Furthermore, they must evaluate the textbook materials' ability to facilitate their teaching of the curriculum standards. Once they have a strong understanding of the textbook and how it aligns with the curriculum, they can then use it to develop their own teaching material in a way that suits the cognitive and psychological development of their students.

While the special role of textbooks is undergoing significant changes (Yan, 2015), the textbook is still the main carrier of the curriculum in China. Furthermore, textbooks also provide students with clues, activities, and structure for their mathematics learning. It is an important tool for teachers when working with students to achieve the curriculum goals. However, recent developments in educational research encourage teachers to critically examine and understand textbooks. Teachers are no longer a passive user of textbooks but are encouraged to further develop the materials in a creative way that suits their own students' development and characteristics. Teachers should examine their teaching materials and organize, adjust, and use them in a way to best allow their students to grasp the mathematics content. In this way, current textbooks have become an important media for teachers' professional growth.

More recently, the Chinese Ministry of Education has largely taken over the development of textbooks. In 2016, several provinces and cities began using the Ministry Textbooks. By 2017, most of the country has switched to using the unified Ministry Textbooks for three subjects: Chinese, Chinese history, and morality and the rule of law. The Ministry of Education is currently working with a group of authors to develop an updated version of the mathematics textbook. By 2021, all schools in

China will be mandated to use the textbooks developed by the Ministry of Education for all subjects. For each subject, there will be a variety of textbooks available. For math, this means that provinces, cities, and schools will be choosing their textbook from a variety of options given to them, but all of them will be published by the Ministry of Education.

3 Use of the Textbooks in Ontario Mathematics Classrooms

Students in Ontario have access to mathematics textbooks and resources from a variety of commercial publishers. These textbooks are created in accordance with the Ontario curriculum and approved by the Ontario Ministry of Education (2005). The approved list of textbooks and resources are published as part of the Trillium List (http://trilliumlist.ca/), which is available online to the public. For elementary mathematics, there are a variety of textbooks and resources that vary from grade to grade as seen in Table 1. The availability of textbooks and resources vary from school to school depending on the decision of the administration and teachers. Most textbooks on the list are large hardcover volumes that are meant to be kept in the classroom from year to year and shared among students. This means that students themselves are not responsible for purchasing their textbooks nor do they get to keep or annotate the books they use.

Table 1 Trillium List (http://trilliumlist.ca/) approved textbooks for Ontario elementary school mathematics

Title Publisher	Grade							
	1	**2**	**3**	**4**	**5**	**6**	**7**	**8**
Addison Wesley Math Makes Sense Pearson Education Canada	■	■	■	■	■	■	■	■
Nelson Mathematics Nelson Education	■	■	■	■	■	■	■	■
Math Now GTK Press	■	■	■					
Mathology Pearson Education Canada	■	■						
Mathematics: Making Connections McGraw-Hill Ryerson							■	■
Math Help Services Interactive Workbook Jitlan Inc.								■

Due to the cost of the textbooks, most schools will buy large sets and use them for many years. Reviewing the textbooks in many of the schools in Ontario will reveal that the textbooks are often multiple decades old and thus not in accordance with the most recent curriculum.

While many schools have textbooks, many teachers interviewed and observed during this project either did not use them or only used them for some of its components. When utilized, the textbooks were often only used for some specific expectations or its questions and exercises. Furthermore, teachers often did not teach the topics in the order dictated by the textbook.

For teachers who do want to use textbooks as the main resource in their teaching, the obstacle they faced most often is the availability of the resource and its appropriateness for the demographic it serves. Many schools, especially schools in lower socioeconomic areas, often do not have enough textbooks for each student to have one. Very often, multiple classes have to share one set of textbooks. Furthermore, recent research by scholars such as Sleeter and Grant (2011) have led to the decreased use of textbooks due to their failure to address the race, gender, class, and ability diversity within the classroom.

The disparity between what is represented within the textbooks and the student demographic in Toronto is large. In certain areas of Toronto, classrooms are made up of non-Caucasian students whom have parents that are first-generation immigrants and thus live within their own cultural traditions at home. In addition to the cultural diversity, the greater Toronto area is home to students that come from a wide socioeconomic spectrum. These students' experiences are not always reflected within the textbooks that are available within their schools. For example, one of the textbooks commonly available gives examples about going on a family road trip and lists a series of expenditures that the students must calculate. In low socioeconomic communities, this trip scenario is often unrelatable to children. The lack of connection the students feel to the textbooks directly impacts their learning as they often will lose interest in the subject. Furthermore, without connection to the mathematics problems presented, students lack the ability to transfer their learning to personal scenarios (Dewey, 1933).

Textbooks have also decreased in popularity within Ontario classrooms as they are often not suitable for students with exceptionalities as they do not differentiate learning, nor do they address the needs for different learning styles. In recent years, instead of following a textbook, teachers in Ontario have opted for lessons designed with "low-floor, high-ceiling" (Boaler, 2014) so that one lesson will address the learning needs of students who are struggling in mathematics and those who need a challenge. Technology in the form of videos and interactive activities have also allowed teachers to venture away from using textbooks as a main resource as they are often more up to date and adaptable to students' specific needs.

4 Classroom Environment and Manipulatives Use in China

In Mainland China, the classroom is the central space for learning. Students spend most of their time in classroom or in schoolyard. The classroom is truly the students' space as the students' different subject teachers have their own offices outside of the classroom. Since most teachers in Mainland China teach one specific subject, they often teach several classes and go to several different classrooms in a day. The classroom is a public place for students and subject teachers to come together. The Mainland China classroom is usually organized so that student students are sitting side by side (Fig. 1a). The advantage of this organization format is that it can effectively use the classroom space

Fig. 1 a. Chinese Classroom b. Chinese teacher using manipulatives while teaching in a Canadian classroom

to accommodate a larger number of students. It is also conducive for teachers to give authoritative lectures in the classroom and effectively monitor classroom activities (Li, 2011).

However, after undergoing a new curriculum reform, Chinese teachers have made many attempts in changing transforming classroom organization. In order to meet students' individual needs, Chinese teachers tried to organize the tables in groups, to encourage and facilitate students to have group discussions in class. Moreover, as constructivist teaching practices are used more often in classroom teaching, people have objected to conventional seating arrangements and explored several new seating arrangements, such as horseshoe, round, and group (Li, 2011). These different kinds of seating arrangements have made some changes to the classroom environment. Teachers are no longer standing on the podium, but are instructors of student self-directed learning (Li & Zhu, 2005). Individual tables that can be organized into multitable groups are often used as they are conducive to promoting student interaction and mutual cooperation. By using individual tables, teachers have the ability to create different seating arrangements according to different teaching purposes and courses.

Manipulatives are widely used in mathematics teaching in China. Teachers in Mainland China use whiteboards and blackboards to assist teaching, as well as PowerPoint presentations to support their teaching. According to students' cognitive characteristics, thinking patterns, and situations stemming from the actual lived lives of students, manipulatives help students better understand mathematical concepts and master mathematical calculation methods so as to cultivate students' sense of innovation (Bai, 2018). Mrs. Wang, who is a Chinese teacher from math team, used manipulatives in a tessellation class that she taught while visiting Canada. She used different shapes of paper to help students understand the tessellation graphics (Fig. 1b).

5 Classroom Environment and Manipulatives Use in Ontario

The elementary classroom in Ontario is structured so that all of the students spend the majority of their time in one classroom with one teacher. The one teacher is a generalist teacher who is in charge of teaching the core subjects, math, language, social studies, science, etc.; thus, the classroom environment reflects the multidimensional needs of the students and teachers. The two classrooms shown in Fig. 2 represent typical elementary classrooms in Toronto, Ontario. As you can see from Fig. 2, the two grade 5 classrooms are very different. The signs, posters, and decoration on the walls not only showcase mathematics knowledge and information but also highlight class schedules, social studies, and STEM. Ontario teachers often try to create a math-positive atmosphere in the classroom, displaying ways in which math is connected to everyday life. Furthermore, teachers will often provide visual aids for students on the walls of the classroom such as number lines and problem-solving process infographics.

Manipulatives are widely used in Ontario classrooms, especially at the younger grades. The availability of manipulatives is largely dependent on school budget and teacher allocation of resources. Some examples of commercially available manipulatives found in Ontario classrooms are shown in Fig. 3. Common manipulatives found in elementary classroom include base ten blocks, counters, and number lines. Other than commercially available manipulatives, teachers will often create their own class sets of manipulatives as seen in Fig. 3.

Fig. 2 Grade 3 classrooms in Toronto, Ontario. As seen in photos above, teachers have autonomy in how they organize their classroom. Walls of the classroom are often used by teachers as a venue for sharing important information with students

Fig. 3 Left: Base ten blocks organized by teacher into kits for student use. Right: Student uses Mathlink cubes as an aid when problem solving

The frequency and the way available manipulatives are used in the classroom largely depend on an individual teacher's math teaching and learning pedagogies. Some teachers may only use and ask students to use a particular manipulative for specific lessons and concepts while other teachers may demonstrate to students how to use a manipulative for a concept and then make it available from then on within the classroom and allowing for students to use at their discretion. The curriculum gives suggestions on how and when teachers can use manipulatives but does not require them to be used in learning.

6 Technology in Chinese Elementary Schools

The *National Medium- and Long-Term Education Reform and Development Plan* (2010–2020) pays special attention to promoting information technology education (Ministry of Education, 2013). This document states:

> Information technology has a revolutionary impact on the development of education, and it must be highly understood in order to strengthen the application of information technology and improve teachers' ability to use information technology. It is essential for educators to update

> teaching concepts, improve teaching methods, and improve teaching effects. Students should be encouraged to use information technology as a means to learn actively, learn independently, and to enhance their ability to use information technology to analyze and solve problems. (Ministry of Education of the People's Republic of China, 2013)

The *Nine-Year Compulsory Education Mathematics Curriculum* (Ministry of Education of the People's Republic of China, 2012) highlights the impact of technology on mathematics education (Wang, 2013). In 2001, Li and Zhang (2001) proposed the idea of an intelligent education platform. They developed a super interactive board that integrated all or part of the functions of various software such as geometric sketchpad (dynamic geometry), PowerPoint (presentation), Excel (spreadsheet), Mathematica (symbol calculation), and Visual Basic (algorithm programming). This interactive board also had dynamic algebra and logic functions such as animation and automatic reasoning, which had played an increasingly important role in mathematics teaching practice in many places (Wang & Yang, 2013).

In today's mathematics classrooms in Mainland China, PowerPoint has become the most commonly used media method to assist teachers in teaching. Courseware and other material designed for use in an educational or training courses now offer teachers a way to show various visuals in mathematics education by using PowerPoint. Most publisher's textbooks have lesson PowerPoints as an accompaniment.

The Chinese Ministry of Education's new requirements for students' core literacy now pays more attention to the overall development of students. Previously, Chinese elementary schools have implemented discipline-specific teaching for teachers, emphasizing students' learning of a single subject (Zhu, 2017). However, currently, curriculum integration and how to train comprehensive students have become new challenges in China's education landscape.

Many schools have begun to explore Science, Technology, Engineering, Arts, and Mathematics (STEAM) education. Research is starting to explore how technology is being integrated into mathematics teaching. The Chinese government will increase financial investment to update the school's technology equipment to provide basic guarantees for teachers and students to improve their ability to use technology. In addition, the use of technology in the teaching classroom is closely related to the level of teachers' abilities, which also puts forward new requirements for the development of teachers' professional abilities.

7 Technology Use in Ontario Schools

A range of technology is used in Ontario classrooms including but not limited to calculators, interactive whiteboards, tablets, and computers. While textbooks are available in almost all schools, the availability of technology greatly varies from district to district, and even between schools within the same district. Some schools in Ontario may have an interactive whiteboard in every classroom and class sets of tablets and laptops to share among classes while some schools simply have a computer lab. This variation largely depends on how schools choose to allocate their funding and the amount of funding received from associations such as the Parent Advisory Council. For students with exceptionalities, there is often additional funding that can be procured to provide necessary technology such as personal devices.

How technology is used in the classroom depends on the teaching plan of the classroom teacher and their comfort with using technology as a teaching tool. To guide teachers, the Ontario curriculum outlines the advantage of using information and communication technology in mathematics teaching and learning. Furthermore, it gives suggestions of when technology could be used to extend and enrich the learning of specific concepts such as geometry and data management, but does not give specific instructions on what to use.

The two main ways information and communication technology is used in the classroom are as a mode to disseminate knowledge to students and as a learning tool for students to explore concepts, practice skills, and showcase their work. Dissemination-focused technologies that are used in the classroom include interactive whiteboards and document cameras. With these devices, teachers are able to show visual representation of abstract concepts, demonstrate using virtual manipulatives that otherwise would not have been available, and differentiate learning for various students.

Student-centric technologies such as calculators, tablets, laptops, and robots are other forms commonly found in Ontario classrooms. These forms of technology allow for further individualized and hands on learning. Furthermore, gamified math programs, online quizzes, and virtual, interactive lessons allow for students to practice newly learned concepts and receive immediate feedback on their understanding in an exciting and safe environment. Ontario teachers also often use student-operated technology programs such a Google Drive to concurrently teach students twenty-first-century skills while facilitating collaboration in mathematics learning.

The combined use of teacher-centric and student-driven technology has become particularly important in a Canadian classroom as inclusive education practices mean that a classroom of 30–35 students may have almost a dozen students with individual education plans that require adapted or modified programs. With special technologies, teachers are better able to concomitantly deliver lessons that are tailored to each student's needs and abilities. In the younger primary grades, teachers find technology to be most useful for its facilitation of interactive exploration of new concepts and repetitive practice with immediate feedback. In the older grades, Ontario teachers most often use technology so students can better visualize abstract concepts.

In recent years, integrated science, technology, engineering, and mathematics (STEM) education has become a popular pedagogy used by many teachers in Ontario. By integrating math with engineering, technology, and science, students are able to gain a deeper understanding of the mathematics content they are learning in class and its application in the twenty-first century (Appelgate, Jackson, Jurgenson & Delaney, 2018). Ontario teachers, especially those teaching in grades 4–8, often use technology such as robotics and coding to facilitate this STEM link. The types of robotics used include but are not limited to Lego Mindstorms, Spheros, and VEX.

8 Teacher as a Resource in Ontario

In Ontario, there is great flexibility and responsibility given to the classroom teacher when it comes to mathematics teaching and learning. From the sequence of curriculum expectation, to textbooks and student activities, the teacher often has complete autonomy in deciding their order and use. While some teachers work together to develop a consistent program within the school, it is not a requirement. When creating a program for the classroom, a teacher often starts by creating a long-term plan for the year that maps out the order that the different strands and topics will be taught throughout the year. Following the long-term plan, the teacher will then create unit plans for each topic that must be taught within the year.

While the Ontario curriculum is divided into five separate strands, there are often multiple large topics within each strand so the teacher will plan according to those topics. This means that each strand is often revisited throughout the year. The topics, otherwise known as unit plans, are detailed plans that each teacher must make, which break down the unit into individual lessons.

When planning a unit, the teacher must use their mathematical knowledge for teaching and pedagogical expertise to create a sequence of lessons (Ball, Hill, & Bass, 2005). It is the teachers' responsibility to evaluate the resources and materials that are used in the teaching and learning of those lessons. Very often, teachers must use a combination of different textbooks and workbooks, Internet resources, and store-bought math manipulatives. In addition to commercially available educational materials, teachers will often create their own resources such as worksheets, test, and manipulatives in order to meet the curriculum expectations and differentiate the lessons for their students' specific needs. This is where the teachers' subject and pedagogical knowledge makes them one of the most important resources in math teaching and learning.

9 Conclusion

In this chapter, we describe the various resources available to teachers of elementary mathematics in Mainland China and Ontario, Canada. There are many choices of textbooks in China. Many large university publishers print textbooks to match their perspective on teacher practice and student learning. The teacher is able to conduct their own research and to develop and integrate teaching activities that fit the characteristics of the students in their region.

The Ontario textbooks are approved by the government (http://trilliumlist.ca/) and are quite old. The Ontario elementary school mathematics curriculum was last updated in 2005 and thus, the textbooks have not been updated recently. These textbooks do not necessarily represent the multicultural population in many Ontario cities. There is high manipulative and technology use in many elementary school classrooms.

In both countries, teacher as a resource is an important complement to curriculum. Teachers prepare their own materials based on the curriculum topics using a variety of sources. While the curriculum is very different in each country, there are many similarities in the resources that can be used in the classroom and the flexibility of the teacher to select appropriate resources.

References

Appelgate, M., Jackson, C., Jurgenson, K., & Delaney, A. (2018). Introducing mathematics concepts using STEM Connections. *Teaching Children Mathematics, 24*(6), 394–397.

Bai, R. (2018). Discussion about how to strengthen the cultivation of students' innovative consciousness in elementary mathematics teaching. *Curriculum Education Research, 5,* 154.

Ball, D. L., Hill, H. C., & Bass, H. (2005). Knowing mathematics for teaching: Who knows mathematics well enough to teach third grade, and how can we decide? *American Educator, 29*(1), 14–17, 20–22, 43–46.

Boaler, J. (2014). The mathematics of hope: Moving from performance to learning in mathematics classrooms. *Heinemann Blog.* Retrieved from https://blog.heinemann.com/the-mathematics-of-hope-moving-from-performance-to-learning-in-mathematics-classrooms.

Dewey, J. (1933). *How we think: A restatement of the relation of reflective thinking to the educative process.* Boston: Henry Holt.

Li, C., & Zhang, J. (2001). Concept and implementation of intelligent knowledge platform. *World Science and Technology Research and Development, 23*(6), 1–6.

Li, L., & Zhu, R. (2005). On the management of classroom teaching environment in primary and middle schools. *Education Management and Evaluation, 8,* 64–66.

Li, R. (2011). The negative effects and suggestions of traditional classroom seating arrangement. *Teaching and Management* (Secondary Edition), 27–29.

Ministry of Education of the People's Republic of China. (2012). *Mathematics curriculum standards for compulsory education (2011 Edition: 4–5).* Beijing: Beijing Normal University Press.

Ministry of Education of the People's Republic of China. (2013). *National medium and long-term education reform and development program* (*2010–2020*) [DB/OL]. http://ishare.iask.sina.com.cn/f/23136097.html.

Ma, Y. (2012). *Primary school mathematics curriculum standards and textbook research.* Higher Education Press, 107–129.

Ontario Ministry of Education. (2005). *The Ontario Curriculum 1–8: Mathematics.* Ontario: Queen's Printer for Ontario.

Sleeter, C. E., & Grant, C. A. (2011). Race, class, gender and disability in current textbooks. In E. F. Provenzo & A. N. Shaver (Eds.), *The textbook discourse: Sociocultural dimensions of American schoolbooks.* New York, NY: Routledge.

Wang, G., & Yang R. (2013). Evaluation standards of mathematics teaching design integrated with information technology. *China Audio-Visual Education* (11), 106–109 + 125.

Yan, J. (2015). *An empirical study on the relationship between "teacher use textbook level" and classroom teaching effect: Taking elementary mathematics as an example.* Northeast Normal University Press, 1.

Zhu, Y. (2017). Practical research on expanded curriculum of primary school mathematics under STEAM concept. *Curriculum Teaching Research, 0*(10), 88–94.

PART II

Reciprocal Learning Between Teacher Pairs

CHAPTER 5

Mrs. Smith and Mr. Jun's Reciprocal Learning Partnership

Sijia Cynthia Zhu

1 Introduction

This chapter shares the journey of two teachers, one from Changchun, China and one from Ontario, Canada as they engaged in a reciprocal learning partnership (RLP). For three years, Mr. Jun and Mrs. Smith worked together in collaboration with University of Toronto and Northeast Normal University researchers, to learn from and with each other on the topics of elementary mathematics. As part of Xu and Connelly's (2013) SSHRC-funded Reciprocal learning in teacher education and school education between Canada and China Partnership Project, the reciprocal learning partnership of the two teachers aimed to build knowledge and understanding across cultures and the discovery of new approaches to mathematics teaching and learning.

When considering teacher knowledge and cross-cultural collaboration between Canada and China, we must consider that education in

S. C. Zhu (✉)
Ontario Institute for Studies in Education, University of Toronto, Toronto, ON, Canada
e-mail: cyn.zhu@mail.utoronto.ca

S. C. Zhu et al. (eds.), *Reciprocal Learning for Cross-Cultural Mathematics Education*, Intercultural Reciprocal Learning in Chinese and Western Education,
https://doi.org/10.1007/978-3-030-56838-2_5

Western and Eastern worlds have developed through extremely "contrasting historical narratives" (Hall & Ames, 1999, p. 11). As part of the Eastern world, China's current communitarian democracy combines ancient Confucian philosophies with Western ideas of individual rights (Hayhoe, 2008). As part of the West, Canada's liberal rights-based ideals are very much at odds with China's current society (Anderson-Levitt, 2008). While the difference is vast, increased globalization will increase interactions between the East and the West, which means that both cultures will inevitably change and move toward each other (Hayhoe, 2008; Xu, 2011).

In terms of comparative education in the hopes of learning and developing educational best practices, we must first situate existing educational practices in their dramatically different Eastern and Western contexts (Sadler, 1979; Xu, 2011). Historian and educationalist Sir Michael Ernest Sadler (1979) stated that:

> Education is that great aggregate of influences which comes to us in our homes, at church or chapel, in daily life, in intercourse with our contemporaries, in love of home and father and mother, in all the thousand streams of influence and suggestion which in a free country converge upon each individual life and shape ideals of conduct. (p. 49)

This definition of education can be seen as the beginning of a cultural context framework when applied to comparative education research and studying the teaching and learning practices of other nations (Beech, 2009).

Sadler (1979) suggests that, when we explore different systems of education, we must be careful not to pick and choose components and transplant them into our systems without careful study of how they work in their own cultural contexts. Therefore, by combining the knowledge about the cultural histories and traditions with current educational practices in the East and West, Hayhoe (2008) believes that we will be able to move into a new phase of more open and mutually transformative dialogue.

The importance of culture, context, and personal experiences is one that is interesting to consider when thinking about mathematics, a discipline that has historically been seen as a language of numbers and algorithms that are universally the same regardless of language or culture. For example, pi always begins as 3.1415. However, when considering mathematics education, Kilpatrick (2013) also suggests that we must consider the history and culture in which it is embedded. Furthermore, since "schooling operates locally, and education policy is made at the local or national level but not beyond… mathematics education always ranges between universality and singularity" (Kilpatrick, 2013, p. 791).

An international dimension of mathematics education was marked by the establishment of the International Commission on Mathematics Instruction in 1908 (Karp, 2013). Karp (2013) and Schubring (2003) argue that, while there was the movement of approaches to mathematics education across national boundaries, there was little research done in the field until the twentieth century. Brickman (1960) stated that, at that time, the international education community began to recognize certain fundamental problems in mathematics education that were shared between many countries even though their education systems greatly differed. Furthermore, through the twentieth century wars and decolonization, there was a development of new states and thus new education systems. Furthermore, alliances between nations during these times spurred the influence of different education systems on one another (Karp, 2013).

From 1980 to present day, there has been an increase in the exchange of people and ideas due to the opening up of borders and the technology revolution (Karp, 2013). Since then, many scholars have completed studies comparing various components of mathematics education in one nation to another (Ma, 1999). Furthermore, there has been a sharing of resources from developed nations such as the United States, Canada, Japan, and Britain to developing nations such as China and El Salvador (Atweh & Clarkson, 2001; Zhu & Monroe, 1991). In more recent years, there have also been collaborations where countries such as the United States, Canada, Britain, and China have strived to learn from each other through pre-service teacher education exchanges and the borrowing of teaching and learning resources (An, 2008; Haas & Weale, 2017; Huang, 2017; Xu, Chen & Huang, 2015).

When searching for research that involves comparative studies in mathematics between Canada and China, there is surprisingly very little. Most research to date draws comparison between the United States and China. Studies involving Canada focus on comparisons with the United States, Australia, and Japan (Ram, 2006). Most of these studies set their attention on discovering the differences in mathematics education and evaluating the differences for the positive and negative impacts (Cai, Ding, &

Wang, 2013; Ma, 1999; Miller, Kelly, & Zhou, 2005). Some examples of these studies include Ma's (1999) study comparing the knowledge of elementary mathematics teachers in China and the United States.

Ma's (1999) research revealed that Chinese mathematics education benefits from teachers with a profound understanding of fundamental mathematics, which not only results in expertise in mathematics but also improves their ability to communicate that expertise to their students. Furthermore, Ma (1999) suggested that the teaching day scheduling for Chinese teachers were beneficial to their professional development as they had more time to study their teaching materials and work with students in need.

Research by Cai et al. (2013), Miller et al. (2005), and Xu (2016) reveal differences and similarities between Chinese and US mathematics education on topics such as instructional coherence, use of teaching materials, and mathematical development in early childhood. Cai et al. (2013) suggest that US teachers focus more on the connectedness between teaching activities, topics, and lessons, while Chinese teachers focus more on interconnected nature of mathematical knowledge, transitional language when teaching and dealing with emerging events, and student thinking. Miller et al. (2005) analyze the differences in mathematics development in preschool-aged children in China and the United States showing that, by the time children enter school, Chinese children are ahead in math skills due to factors such as cultural values and the language of arithmetic. Xu (2016) expands on this knowledge by comparing North America with China in terms of using math stories when teaching mathematics.

Building upon those comparative studies, An (2008) and Xu et al. (2015) wrote about pre-service teacher exchange programs and the opportunities for Canadian and Chinese pre-service teachers to engage in short-term practicum experiences. These experiences allow these teachers to learn from another culture/s teaching and learning practices in a variety of subjects including mathematics. Huang (2017) states that, by having embedded cross-cultural experience, new teachers can acquire new educational knowledge that transform their prior educational practices and beliefs and create new integrated pedagogical practices.

While there has been considerable comparative research done comparing different countries, there has been little work focusing specifically on Canadian and Chinese education and even more rare are studies that actively involve the teachers and students from different countries working together as peers. Furthermore, cross-cultural collaborations will link societal issues with mathematics education (Singh & Ellerton, 2013). Xu and Connelly's (2013) work with creating sister school networks and long-lasting collaborative relationships at the University, school-district, school, administrator, teacher and student levels is the beginning of the type of cross-cultural research that authentically involves participants and that are guided by their learning needs.

2 Methods

In studying Mr. Jun and Mrs. Smith's reciprocal learning partnership (RLP), we used case study methods to structure the research design (Stake, 1995). Within that model, we employ a narrative lens (inspired by Dr. Xu and Dr. Connelly) to guide our thinking throughout the data collection, analysis, and discussions phases (Merriam, 1998). As an extension of the larger Canada–China Project, our research attempts to stay true to their overall goals and employs a narrative inquiry framework, which is grounded in human experiences and how they are affected by personal and cultural history (Clandinin, 2007; Connelly & Clandinin, 1990; Xu, 2011). This research also follows the purpose of narrative inquiry as it does not seek to summarize and report the experiences and observation collected, but rather to use them to pose further questions for follow-up research and eventual interpretations. Through using qualitative methods such as observations, conversational interviews, and other descriptors of experience, we were able to gain insight into the development of the reciprocal learning story of Mr. Jun and Mrs. Smith as well as discover cross-cultural pedagogical strategies that are beneficial to mathematics education.

Each case, otherwise known as a teacher pair, will be looked at as a system and, as researchers, our job has been to get to know the pair well and learn "what it is, what it does" without the need to discover how it is different from others (Stake, 1995, p. 8). The use of case studies works well with Connelly and Xu's (2015) narrative vision for the partnership project as it allows for the collection of a multitude of data types ranging from observations, interviews (formal and informal) to documents (Stake, 1995; Yazan, 2015). With the evidence, we have been able to build a narrative for each teacher that focuses on the personal and cultural histories as well as the experiences of the teacher pair. The data collected will not be seen as answers or solutions but instead they will be used as paths for further inquiry and then eventual interpretation (Connelly & Clandinin, 2006; Xu & Connelly, 2010).

3 Getting to Know Mr. Jun

Mr. Jun went to the Changchun Normal University in Changchun, China and majored in mathematics. He had chosen to pursue the teaching profession as he believed that teachers are respected and because of a past teacher that he looked up to. After graduating from university in 2004, he started teaching at the Primary School Attached to Northeast Normal University (PNENU) and has taught mathematics there for the past 10 years. During his time at PNENU, Mr. Jun also went back to school at NENU and earned a Master of Education degree.

Over the 10 years, he has taught grades three, four, five, and six mathematics. During the span of the project, Mr. Jun taught grades four, five, and six mathematics. As teachers in China often follow their students as they advance from grade to grade, Mr. Jun taught mathematics to the same two classes of students as they went from grade four up to primary school graduation in grade 6.

As PNENU had 10 classes of students graduating primary school in 2017, Mr. Jun was part of a teaching team of five math teachers, each teaching two classes. Mrs. Yang, another math teacher on this project, is one of the teachers on that teaching team. When asked about the math teaching team, Mr. Jun says that being part of a teaching team decreases his workload for things such as writing exams and creating teaching slides, which frees up more time so he can meticulously prepare the lessons he is responsible for.

The teaching team also gives him peers to investigate and discuss lessons with, leaving him with a better grasp on lessons. In addition, when issues arise in class regarding the content or pedagogy, he is able to share them with the team and work on a solution together. However, being part of a structured teaching team also has its disadvantages. The team tries to teach the same content and at the same pace, which allows for little flexibility in his own lessons. Furthermore, because exams are developed by the team, he feels like he is not able to create questions as he would prefer to do.

Mr. Jun's main goal over the past ten years teaching mathematics has been to foster a love for mathematics in his students. He believes that this is particularly important for this age group because the grade four, five, and six students have developed increased emotional intelligence. With heightened emotional intelligence, students require interactive lessons where they can communicate their thinking and learning with their teachers and each other.

Conceptually, Mr. Jun believes that fractions are the most difficult topic for his students to grasp. More specifically, word problems involving fractions are especially difficult. He believes that the difficulty lies in the fact that these problems require many pieces of math knowledge to solve. Additionally, a common approach to word problems is through drawing diagrams, which is difficult when it comes to fraction word problems. However, drawing is still a good method for students to investigate the relationship between the fractions.

Whether it be intuitive drawings, schematics, or rough sketches, diagrams are good way for students to visualize the problem. Drawing also helps students map out their thinking process. He says that drawing leads to deeper thinking and problem solving while the sharing of diagrams leads to better communication and reflection. However, he says that drawing diagrams for fraction problems is often difficult. He also says that it is very important in fraction word problems for students to figure out what is being compared to what. For example, students must figure out if the question is asking for them to compare A to B or B to A because the fraction that results is different. Lastly, he believes that there are many types of fraction word problems and can be confusing for students.

When it comes to professional development, Mr. Jun takes part in yearly sessions that occur online. These sessions are organized by the city of Changchun and mandatory for all teachers. In addition to online sessions, Mr. Jun also takes part in school professional development that is put on by educational researchers, teachers, and administrators within the school.

He has been selected multiple times to represent his school in teaching competitions. In October of 2016, he won an award for teaching at the Jilin Province Teaching Competition. Experienced and exemplary teachers at PNENU are often selected to represent the school in competitions and special projects in order to offer them a challenge, and give them an opportunity at new learning that they can bring back to their peers. It is in this way that Mr. Jun came to be involved in the reciprocal learning partnership in 2014. Selected by his Vice Principal, Ms. Wang, Mr. Jun was one of two teachers on his teaching team to join the project as a reciprocal learning partner.

In addition to teaching mathematics, Mr. Jun also leads one of the many clubs at PNENU. His favorite club is the Flying Cups Club. Flying cups is a competitive game that is played globally. Mr. Jun finds that the students who choose his club are often very social and competitive.

4 Getting to Know Mrs. Smith

Mrs. Smith went to York University in Toronto, where she obtained an Honors Bachelor of Arts in political science with a minor in history degree. Later in life, she went back to York University and completed a Bachelor of Education degree in intermediate/senior history. When asked why she became a teacher, Mrs. Smith says:

> I became a teacher after having worked in a volunteer teaching capacity within my community. It began with adult education then eventually I began to teach children of all ages. As time passed, I realized that I loved teaching and learning which resulted in me returning to University for the Bachelor of Education degree. (interview, April 11, 2014)

Her passion for all types of learners was immediately evident in that the first teaching assignment she took on out of teacher's college was as a grade 6 special education teacher. She spent five years at her first school and specialized in teaching students with exceptionalities, specifically students on the Autism spectrum between grades five and eight.

I first met Mrs. Smith during a professional development session hosted at the Ontario Institute for Studies in Education on middle school mathematics. Her enthusiasm toward learning was immediately evident when she raised her hands with questions or comments at every given opportunity. At that session, she was representing her school with five other colleagues. When looking for participants for this project, I approached the math teachers from that session through email and she immediately responded with interest. During the course of the project, Mrs. Smith taught grades six (2014–2015) and eight (2015–2016) at her school and grades 3–5 special education (2016–2017) at another school.

The school where Mrs. Smith was teaching at the beginning of the project is a school where students face many challenges. Mrs. Smith describes that many of the students that attend her school do not have the money to pursue higher education or have the support at home when it comes to schooling. She believes that:

> Character development is a big, huge success for me. Story and kids falling in love with learning is my success. Literacy yes, numeracy yes, but the fact that they actually fall in love with learning is where I find my success. If I accomplish that by the end of the year and my kids are engaged and are like just absorbed in the learning. That is success for me. (interview, April 8, 2015)

In order to help her students fall in love with learning, Mrs. Smith utilizes all of the resources she can get her hands on, even if it is out of date. Mrs. Smith believes that the lack of resources at her school is the biggest barrier to her teaching. Mrs. Smith says that:

> I have never been at a school like this, it broke my heart at the beginning of the school year when I realized that we did not have enough books for the kids. We are talking about textbooks and I know that we do not teach by textbook, but I think that it does something to the child when they cannot have one, you know and everyone else that they know does have one... I think that the resources, the technology is not available here and I know that this is my first year but this is the first time we have a computer lab so I know that our current principal has worked very hard at achieving that... Finances are definitely required here. Whereas other schools have fundraising opportunities, but in our community, we do not have the families with the budget then the fundraising will not be there. (interview, April 2, 2015)

When asked specifically what her goals are for her students in science, technology engineering, and mathematics (STEM), Mrs. Smith focuses her attention on understanding mathematics and helping students persevere even when they are struggling. As someone who "dropped math as fast as [she] could" in grade 11, Mrs. Smith attributes much of her struggles and previous attitudes toward math to "teachers that made [her] hate math" and because of that reason, she is very careful as a teacher to make sure her students do not develop bad attitudes toward the subject. She states that:

> I encourage my kids to not drop math whatever you do, even if you are struggling, even if you need to take extracurricular things or take math tutoring or whatever...do not drop math! Because it is applied to so many areas of our lives and you do not know what you want to do at the end and it might mean that you might need the math. (interview, April 2, 2015)

Mrs. Smith also believes that strong foundational skills are essential to success in mathematics. She feels that, very often, the curriculum forces students and teachers to move onto new content before they have mastered the basics such as simple multiplication. Additionally, due to her background in special education, Mrs. Smith feels that it is important to see students as individual learners and to diversify lessons to meet the needs of each child. But most importantly, Mrs. Smith emphasizes that a positive learning environment is essential because, if the students are not happy in their learning environment or do not feel respected, they are not learning.

During an interview, Mrs. Smith shares that her relationships with fellow teachers are essential to her feeling supported, motivated, and engaged in her work. Mrs. Smith described her peers as selfless, often working until 10 o'clock at night. Teachers teaching the same grade also met weekly to discuss students' learning; these meeting often occur during lunch break.

In addition to working with colleagues, Mrs. Smith also participates in professional development in the form of workshops and professional learning communities that are organized by the school district and school. In her own time, Mrs. Smith also seeks out further opportunities for learning such as participating in research projects such as this one and taking additional qualification courses in special education and mathematics.

5 Learning About and from Each Other

Mrs. Smith and Mr. Jun's RLP started in April 2015. Using a combination of synchronous and asynchronous communication methods, the teacher pairs first learned about each other. Haythornthwaite (2002) outlined that strong collaborative relationships require partners to be bonded in both a personal and professional level and thus the process of learning about each other was designed so that the teachers would be able to access each other on both these levels. This chapter will specifically focus on what the two teachers learned about and from each other regarding mathematics teaching and learning and how it affected their own practice.

6 Learning Over Synchronous Video Chat

At the beginning of the RLP journey, Mr. Jun and Mrs. Smith primarily communicated with each other through synchronous video chat with the help of translators on both sides. Over the course of the first two years of their partnerships, the teachers engaged in six synchronous video meetings over Skype. These meeting largely focused on the teachers getting to know each other, learning about each other's teaching environment and practices, as well as exploring areas of common interest on the topic of mathematics education and professional development. For each meeting, an agenda was developed by researchers in collaboration with the teachers to focus on communication. The agenda focused on topics that the teachers had expressed interest in either through previous video calls, interviews, or conversations with the researchers.

Each video chat provided ample time for the teachers to engage in social interactions and chat about current pondering in regard to mathematics teaching. Over the course of six video meetings, the teacher explored areas including but not limited to: Chinese and Ontario mathematics curriculum, integrating mathematics with other school subjects, differentiated teaching, mathematics teaching and learning for students with exceptionalities, and learning about geometry and fractions.

During the first Skype meeting, the teachers discussed two main topics—curriculum and student interactions between classes. Although there was an agenda, the teachers felt free to deviate from the plan and asked each other questions as they came up. For example, the Mr. Jun asked Mrs. Smith a question about the EQAO test that she brought up in one of her answers, wondering if review for the assessment was separate from her regular math lessons. Furthermore, the two teachers delved into a conversation that compared their two curriculum structures without prompting.

Their conversations revealed that both the Chinese and Ontario math curricula are structured into "strands" or topics that are consistent from year to year. Mrs. Smith describes that the Ontario curriculum is comprised of five strands and gives examples detailing each. Mr. Jun believes that the Chinese curriculum is similarly structured to have four strands. While both curricula have strands focusing on numbers, algebra, geometry, and statistics, the Ontario curriculum breaks down strands such as geometry into a more detailed measurement and geometry and spatial sense strands. Furthermore, the Ontario curriculum also separates number and algebra by having two strands named number sense and numeration and another named patterning and algebra. In the Chinese curriculum, there is also the additional strand titled comprehensive practice that is not found in the Ontario curriculum.

The teachers also spent about half of their meeting talking about their students' participation. Mrs. Smith shared:

> I actually went on Internet and found the video of your school and I showed it to them. They were just overwhelmed actually by the number of students because our school are not as large as yours. And a lot of my particular students, I do not think outside of their particular culture and then Canada so my goal in doing this exchange is to open up a global classroom. (video conference, July 2, 2015)

Mr. Jun said that he too felt it was important for his students to take part in activities such as this. It helps them improve their English and helps them gain a better understanding of Canadian culture and learning. He also added that this project has encouraged him to practice his English and use it in his teaching of mathematics since it is not only exciting for his students to hear their teacher practicing English, but also useful for them to learn the vocabulary in English too.

During a subsequent video meeting, Mrs. Smith discovered that both she and Mr. Jun were teaching fractions. Mrs. Smith was just finishing up with the unit while Mr. Jun was just starting. Mrs. Smith shared that she believed that they would work well together because her students struggled with fraction concepts, so she was looking for more strategies for teaching this topic. Both Mrs. Smith and Mr. Jun shared resources and pedagogical strategies. At one point during the conversation, Mr. Jun explained to Mrs. Smith that, in Chinese elementary math education, the topic of fractions is seen to be one of the toughest topics for teachers to teach and for students to understand. Mr. Jun told Mrs. Smith that usually he only spends a few classes on each topic but for fractions, he takes almost a whole month. This prompted Mrs. Smith to reconsider how long she spends on fractions. She decided that she would take more time later on to review the topic.

In addition to sharing information about their own mathematics teaching and learning from the other, the two teachers also began their planning of a joint activity they had discussed trying out over WeChat. Mrs. Smith recommended that both classes do an activity that connects geometry to engineering and building. Her own class was studying the Pythagorean theorem so she was going to ask them to build a kite and write a report using math vocabulary that shared their design, measurements, and linking Pythagorean theorem to the kite. Mr. Jun thought that Mrs. Smith's activity was very appropriate and highly integrated.

Mr. Jun shared that he had also asked his students to build structures when teaching about triangles. His lesson asked students to build a bridge using toothpicks and using triangles for structural strength. Mrs. Smith was excited to hear that her partner had experience integrating building and math and asked him to share his plans and experiences as she was doing her lesson for the first time. The two teachers exchanged emails and agreed on their lessons plans. While neither teacher used the kite building lesson in their teaching that semester, both were motivated to add hands-on activities to their lessons (see Fig. 1).

During Mr. Jun and Mrs. Smith's RLP, Mrs. Smith began teaching at a different school in a new role as a special education teacher in a class for students with autism. This new role prompted the two teachers to discuss mathematics teaching and learning for students with exceptionalities. Mr. Jun brings up that the concept of special education classes within a regular school is very new to him. Furthermore, while he does have a few students with autism in his classes, they do not have individual education plans

Fig. 1 Mr. J and Mrs. N communicate using WeChat about teaching geometry using hands-on activities. Researchers help translate the teacher's messages

tailored to their needs. Mrs. Smith was able to provide him with some modification and accommodations that she uses with her students.

Both teachers believe that group work allows for students with autism to bond with their classmates and provides them with more "little teachers." It is important to create a community where students take pride in their group and aim to help each other succeed. Furthermore, special education students should be given opportunities to communicate their learning to the class even though it is not in line with the learning of the rest of the class. Both teachers felt that it is important to create specific questions for students with exceptionalities that require them to use their learning but are not outside of their capabilities so that they are able to build their confidence.

As the RLP progressed, the video calls became more focused on the teacher's mathematics interests and the teacher interactions became more at ease. On a video call in November 2016, the teachers began the conversation like old friends, quickly catching up on how the other had spent their day. Mr. Jun shared that his day had been very busy with him teaching two classes and also observing two classes. This prompted a discussion about lesson observation as a means of learning for teachers.

Mr. Jun shared that, very often during the semester, teachers have the opportunity to observe another teacher and vice versa. Their classrooms are thought of as open spaces where researchers, parents, administrators, and other teachers should be able to learn from each other. Teachers are always given notice of these visits, but they happen often. Mrs. Smith felt that this was a great learning opportunity for teachers. She shared that, in her culture, teachers are simply not used to it. She describes:

> The culture here is that you are in your class, nobody comes in that does not belong in there, except for the kids. I think they are self-conscious. A lot of the teachers are self-conscious of being judged or being assessed. So there is not this concept of that when you come in you are just observing to learn. Even yesterday with my mentoring, they specifically pointed to the fact that it's not an assessment from the mentor's perspective. It is just fear, fear.

She also says:

> But that is opening up here now. Because a lot of where we are going is a lot more open doors, and observations, model classrooms. So we are

> going in that direction, but it seems to be more of the younger teachers that are coming in, the newer teachers. Not young in age, but young in experience. And they are not afraid of that. Whereas the ones who have become accustom to not having people in their class, are very apprehensive.

These examples of self-reflection show that Mrs. Smith is acknowledging the differences between teaching culture in the two countries but also reflecting upon why her culture is the way it is. Further conversation between the teachers lead to a discussion on different ways teachers can observe each other and collaborate without fear. Both teachers agree that, when teachers are watching other teachers teach, it should not be about picking out problems. It is more about discussing, "like how could we teach this?, what about this course, how could we did deeper into this topic? It is more about ideas, and not about blame" (observation notes, March 7, 2017).

In the second portion of their discussion, the teachers shared the topics that they were currently teaching, and while there was no overlap, the two were able to give each other suggestions and share past experiences. Mr. Jun shares that he is teaching percentage word problems and that he finds it to be a very difficult concept for students. He hopes that students will be able to draw upon their learning from last year on the topic of fractions word problems since the two are very similar. Furthermore, he will encourage his students to use diagrams to help them solve the problem.

In response, Mrs. Smith shares that, in her experience, Ontario students usually learn the two concepts together in the same year so they can see the correlation. However, she feels that how the Chinese curriculum puts fraction first and then address percentages may be a better idea. Mrs. Smith shares:

> It goes back to the same discussion we have over and over again in the sense of having the children really established in the fundamental that is necessary to understand all these abstract or other ideas. So, I absolutely agree with him. I think ours is way too much. Because we do ratio, fractions, and percentages all in the same year. (video conference, November 18, 2016)

Later in the conversation, she says:

> My particular pet peeve, about the way our system teaches math, is how can I possibly… if my kids don't even understand what a characteristic is,

the concept of characteristics, how can I now delve into 3D shapes, 2D shapes? And expect them to have a clear understanding of that? (video conference, November 18, 2016)

This example shows that Mrs. Smith is beginning to find a pattern she believes makes for successful learning in China. This is reflected in her own classroom as she spends considerable amount of time that year on foundational knowledge. To her, foundational knowledge is found mainly in the number sense and numeration strand of the Ontario curriculum.

Both teachers believe it is important to connect the math they teach to real-life applications that are relatable to the students. Mr. Jun points out that the textbook questions need to be up to date and specific to where students live. Questions such as “the farmer harvested 200 tons of crops last year but this year he has 50% more…” may make sense to students who live in rural areas, but for city kids, they make little sense and are not fun. So, instead of using agriculture examples, he modified his questions so they are about typing speed.

Mrs. Smith was inspired with that example and reflected upon what that would look like in her own class. For the students with autism, it is even more important that they connect with the math on an even more personal level. She feels that she should modify her questions to suit specific students.

These examples show that teachers from both countries are actively trying to engage their students in context-based and differentiated lessons. Mrs. Smith also shared a rap song about ratios with Mr. Jun since she has noticed in videos that Mr. Jun uses chanting as a teaching tool, and that his students enjoy singing. With my help in translating, Mrs. Smith was able to send a subtitled version of this song to Mr. Jun.

The conversations during the last three Skype sessions are evidently more relaxed. More time is spent on discussing mathematics and making plans for sharing resources than on social and personal conversations. The majority of personal conversations have occurred through the WeChat platform. Mrs. Smith describes that November 2016 experience as “much more natural and fluid. And we actually talked more this time than with the agenda.”

7 Learning Over Asynchronous WeChat

Mrs. Smith and Mr. Jun's WeChat correspondence began because both teachers were looking for more convenient ways to send each other resources. Mr. Jun was already familiar with the technology and Mrs. Smith was able to learn it quickly. Mrs. Smith and Mr. Jun's WeChat group consisted of six people: the two teachers, one researcher on the Canada side, and three researchers on the China side.

Throughout their RLP, Mr. Jun and Mrs. Smith sent each other over 150 messages. The messages contained information about mathematics teaching, students' mathematics learning, teachers' administrative duties, and personal messages. Approximately 70 of the messages focused on mathematics teaching and learning (see Fig. 1), while the rest were dedicated to words of encouragement, sharing about their families and personal lives, and sending each other best wishes on holidays.

Some examples of messages focused on mathematics teaching and learning include:

Mrs. S: I have finished Fractions but the students will be talking about what they learned in their letters to your class. By the way we will be sending them out in few weeks. Some of my students had a difficult time doing various operations using fractions. Maybe you have some strategies I can use to help them understand this math better (WeChat, November 22, 2015).

Mr. J: What great visual demonstrations. Yes, some of my students have difficulty understanding this. I will use your visual models next time (WeChat, December 5, 2015).

Mr. Jun responds saying that, when teaching fractions, he uses strategies to solve the problem such as drawing line drawings to help students sort out ideas, identify the combination of old and new knowledge, and the use of transformation thinking.

When messaging about mathematics via WeChat, the teachers shared their own teaching by taking pictures of their blackboards, textbook pages, and pictures of their students participating in math learning. During the winter holidays, both Mrs. Smith and Mr. Jun shared pictures of hands-on geometry lessons activities that their students had completed, which gave Mrs. Smith the idea of planning a lesson together with Mr. Jun and seeing how it unfolds in each of their classes.

Another example of the teacher's collaboration on WeChat was on January 16, 2016 when Mrs. Smith sent a picture of her students solving a problem on the whiteboard, stating that some students had trouble with the problem. Mr. Jun had taught this problem previously and took the time to quickly create a figure that he has used to help students understand. Mrs. Smith thought it was a great graphic and used it during her next class to help some of her struggling students. The teachers said that WeChat was an effective platform for them to keep up to date with each other's teaching, and ask each other questions about one another's assignments and activities. The teachers also helped one another improve their teaching.

In addition to sharing about their own teaching, the teachers used this WeChat to ask each other for ideas and resources for lesson they were planning. In one case, Mrs. Smith was planning a lesson on measurement and thought of a picture of a lesson that Mr. Jun had shared when he taught estimation of area. She saved the picture and used it for her lesson. She later shared this information through a Skype conversation. Their conversations on WeChat even extended into the summer when Mr. Jun was planning for a teaching competition and asked Mrs. Smith for resources that she used to teach circumference to her grade 8 class.

8 Learning Through Asynchronous Lesson Study

Through discussions with both teachers, it was clear that they were interested in seeing their partner in action at school with students. During a video conference, Mrs. Smith had communicated that she thought watching other teachers was a great opportunity for learning and that she did not get to do it enough in Canada.

Prior to the lesson study, the teachers shared their thoughts on why lesson studies are important, what they watch for while observing a lesson, and what they believe is a successful lesson and effective teaching. During the lesson study, the teachers were asked to verbalize their thoughts. After watching the complete video, the teachers were asked a series of questions.

Multiple videos of the teacher were collected during their RLP. Each teacher was given the opportunity to choose one of their partner's videos to complete the lesson study. Mr. Jun chose to watch a longer and subtitled version of Mrs. Smith's Time lesson as it was most applicable to the grade he was teaching. Mrs. Smith watched a video of Mr. Jun teaching three-digit by two-digit multiplication, which I narrated in real time.

Before watching their videos, both teachers shared that their focus when watching a lesson was on the students since they are the focus and recipient of the lesson. When watching a video, Mrs. Smith asks herself:

> Is the teacher able on the spot to gauge whether the kids are understanding? First of all, how is she gauging that? It could be as easy as observation, right, and interaction. And if the students are not picking it up, what is the teacher willing to do? Is the teacher willing to abandon their lesson plan and go and do something that will make the students understand? (interview, November 21, 2016)

Mr. Jun too states that he focuses his observation on the students to see their reactions to the learning activities, since only when students are willing to participate will the teaching have any meaning.

Both teachers also suggest that teachers need to be flexible and creative when teaching. If an activity does not engage students or is simply not working to help the group of students understand, teachers need to be prepared to make changes on the spot regardless of their plans.

Additionally, Mrs. Smith and Mr. Jun state that it is important that lessons help students understand the concepts being taught instead of just mimicking the rules. This requires teachers to design activities that push students to use their higher-level thinking.

Mr. Jun expands on this by saying that, while activities should be stimulating and exciting, their design needs to be simple and congruent with student abilities so that learning feels comfortable. Mrs. Smith believes that it is much easier for subject specialists such as Mr. Jun to be flexible and creative in the math classroom since he has a better understanding of the content. She shares an experience in class when a student brought up an interesting question that was not within her planned lesson and while she was able to address the question later, she felt badly about missing out on a teachable moment.

Mrs. Smith watched Mr. Jun's lesson on three-digit by two-digit multiplication on November 21, 2016 with me translating the video as she watched. While she watched the video, she focused on the language he used to break down the concept he was teaching. In her own words:

> the way that he introduced the question by having this abstract idea. Okay, this has nothing to do with math, and then it is something that engages the student, right. So that was really interesting. And the idea of letting the kids pull out the math by pulling out the number in that question and turning it into a math problem. And so I liked that, and the other thing I liked was the way that he questioned the students. Because I personally have a hard time not jumping to the answer. But allowing my questions to be those probing things that allow the kids to come up with the ideas. I learned a lot about that. (interview, November 21, 2016)

Mrs. Smith also paid attention to how Mr. Jun arranged his lesson to incorporate group work and how he navigated the logistics. This topic of grouping was a consistent topic for the two teachers so she was happy to see it in action because it was easier to understand visually.

In addition to pedagogical learning, Mrs. Smith also felt that she was able to improve on her mathematics content knowledge. During the interview, she shared that:

> I am going to tell on myself now. In the video where I had this "awe" moment, "oh now I understand why we put that zero there!" Because when I was learning math, I do not think my teacher ever… they just said put a zero there, and never really explained, oh that is a place holder for ones column. (Mrs. Smith, Interview, November 21, 2016)

Overall, Mrs. Smith enjoyed the lesson very much and found it to be a good learning experience, saying, "I love the fact that it was videotaped because it gives me a visual, and I remember more through visuals. The pacing of the question, and the time between the answers" (Mrs. Smith, Interview, November 21, 2016). Furthermore, she stated that she would use examples and questioning technique that Mr. Jun used in her own teaching.

When asked if she had any questions or suggestions for Mr. Jun, Mrs. Smith added that she would recommend asking the students to add a final conclusion sentence that requires them to articulate their conclusion in words. Furthermore, she wondered about how a teacher should arrange groups to answer questions. Should they start with a group that has done the activity in the "proper" way or should we build up to the "proper way"?

When watching Mrs. Smith's lessons, Mr. Jun focused primarily on how Mrs. Smith communicated with her students. Mr. Jun noted that

Mrs. Smith had a very natural way of interacting with students and responding to students' answers and questions. Mr. Jun also observed that Mrs. Smith gave many visual cues to her students, such as an outline of her lesson and learning goals for the class, which allowed for students to stay on track and immediately interact with the teachers and keep the lesson on track.

While he thought it was very interesting and useful that Mrs. Smith connected time zones to the 24-hour clock, he wondered if it was too confusing for students to learn together. Mr. Jun points out that, while the topic and introduction to the topic of 24-hour clock is similar, Mrs. Smith covers a wider breadth by also teaching the students about time zones while his version of the lesson would focus on the 12-hour and 24-hour clocks.

In addition to answering interview questions on the video observation, Mr. Jun also took the initiative to adapt Mrs. Smith's lesson and write a script of how he would communicate with students about this topic. This exercise was meant for his own personal learning but also for Mrs. Smith as she had previously expressed the desire to learn from Mr. Jun in how he posed questions students when teaching. This was translated and sent to Mrs. Smith who was excited to use it in her own class and touched by Mr. Jun's efforts and thoughtfulness.

The teacher's interviews were shared with their partner. Reading someone else's perspectives and observations of their teaching, especially ones from a teacher they had a long-term relationship with, was very motivating for the teachers. Both teachers stated that their partners had noticed positive things about their teaching that they had never noticed or thought of as an effective strategy. Upon reading their partners' interviews, the teachers responded to questions and inquiries. I translated their responses. Even though the teachers had seen video of each other, and talked extensively over video conferencing, having the time to go through a video lesson in detail brought up new ideas for collaboration and study.

9 Teacher Exchange and Learning Together

During the conversations over Skype, WeChat, and PeppeR, it was evident that the two teachers wanted to meet each other. Mrs. Smith often said how it would be so interesting to see Mr. Jun teach in his class. Mr. Jun echoed this sentiment saying that it would be great to see how Mrs. Smith was able to work with all the different students in her class and manage the integrated lessons. As the teachers continued their correspondence, it was clear that many of their questions could be much easier answered through visiting each other and experiencing each other's teaching environment in situ. Thus, after three years of long-distance RLP, Mrs. Smith and Mr. Jun traveled to each other's classrooms to work with each other in March and May of 2017, respectively.

9.1 March 2017

Working together with Mr. Jun and his Vice Principal Ms. W, Mrs. Smith developed a lesson that would be suitable for his class. At the beginning of the exchange planning, the teachers had hoped to work together to create a joint lesson that would be taught in both classes; however, it was evident that Mrs. Smith's grade 3–5 special education class and Mr. Jun's grade 6 class had very different needs and it would be too difficult to develop one topic/class that would suit both.

Through discussions, and looking through Mr. Jun's curriculum topics for that semester, Mrs. Smith found that the circle, circumference, and cylinders was a big topic in the grade 6 curriculum. This prompted Mrs. Smith to choose a language and math integrated lesson for Mr. Jun's class that highlights the topic of circumference and pi. Mrs. Smith chose this lesson because she felt that this integrated way of teaching would be something new for the teachers and students in Changchun.

Integrating language and math is also one of the ways that she has worked to make math approachable to her students and herself. Furthermore, Mrs. Smith decided to bring her own manipulatives to use for her lesson that were symbols of Canada so that she could leave them with the students as souvenirs. In addition to planning for her lessons, Mrs. Smith also prepared resources and gifts for the teachers.

Mrs. Smith's trip to Changchun was only four days so her schedule was tightly packed with activities. On the first day of her visit, Mrs. Smith shadowed Mr. Jun's daily teaching life through attending his classes, discussing with him his assessment processes as well as participating in his grade 6 math team lesson preparation meetings.

During the observation and participation period of her visit, Mrs. Smith observed that:

> The kids seemed to be very comfortable having other adults in the class and did their activity as usual. Mr Jun taught a three-part lesson plan.

> Proportional distribution, books vs toy cars. The math focused on one question and lesson was very specific students didn't have to deal with a lot of different information. This was very interesting because I was expecting a lot of material in a route model. Instead it was interactive, fun and focused. Students seemed very accustomed with the process and went through it very naturally. Each group worked well together, each student participating and sharing. (Mrs. Smith, March 9, 2017)

This is an example of a point that Mrs. Smith highlighted repeatedly throughout her visit. She had previously thought that the students would be doing more rote learning and simply practicing calculations; however, she found the students were participating in group work and inquiry-based lesson.

She found that the only "rote" part of the lessons was that the students had a strong grasp of the process of learning techniques such as group work, inquiry, presentations, and gallery walks. She feels that this is very important part of a good math lesson since the students were able to efficiently and effectively use class time while consistently feeling confident about their learning. Through her discussion with the teachers, she learned that the classes start teaching this learning routine as early as grade one.

Mrs. Smith found the math team preparation meeting to be very interesting and not what she expected. She had initially thought that the process would be very formal and dry but she found the atmosphere to be very open. She pointed out that:

> There was a lot of collaboration that was very free and open, and no judgement, and so they were colleagues but no one had to prove whether they were right or wrong. They shared the leadership. And even the lead of the department, she was so interested in the way she played the devil's advocate but not in a devilish way or in a way of condescension. (March 9, 2017)

She believes an essential part of what makes the teachers' collaborations successful is that they focus on the students instead of who has the better teaching methods. Mrs. Smith feels like the teachers focused on examining the lesson from the perspective of the student. Furthermore, during the meeting, Mrs. Smith observed how the teachers used the pre-determined textbook and teacher PowerPoint slides, and found it interesting that they did not see it as a restriction but as a foundation that they can modify one. She felt that this affected her view of textbooks and hoped that she could use them similarly in Canada.

Prior to the trip, Mrs. Smith was nervous to be teaching in Mr. Jun's class because she felt that Mr. Jun was an exemplary math teacher and she would not be able to teach as well. However, after observing Mr. Jun's class in situ, she began to feel much more comfortable. Overall, she felt that her lesson had gone well, and that the students "were engaged and got it!" and that the Canadian cultural content embedded in the activity was fun.

During her lesson, Mrs. Smith had to use technology that she had not used before. While she was initially nervous and "almost scrapped it during the lesson," she felt good about challenging herself and using it till the end. She realized that her fear with technology was about herself and not about the students.

Mrs. Smith points out how easily the students adapted their regular learning routines to her teaching and her tasks. She describes that, once she gave her instructions, the students were able to quickly get on task even though it was a different teacher and totally different lesson. Furthermore, she found that the students were confident in asking questions and presenting their work to her and all of the other teachers in the room. She attributes much of that to their confidence in their learning routines.

After teaching the lesson, Mrs. Smith participated in the post teaching discussion with teachers where she found that the "teachers all had good questions and were interested in using the strategy in their math lessons." She was initially afraid that they would ask her math questions that she could not answer but was relieved when most of their questions were focused on pedagogy. Furthermore, she noticed that the teacher was very supportive and did not focus on what did not work, but instead they asked questions that would allow them to apply the new pedagogy to their own classrooms. Mrs. Smith pointed out that the teachers were very quick to take action, "brainstorming Chinese literacy books that they could use."

Upon leaving Changchun, Mrs. Smith reflected on her trip saying that:

> I felt like we knew each other because we had spent so much time with each other, it was not actually a lot of time, but the fact that we did interact, and we not only interacted as colleagues around math professionally but we asked about each other's families and we cared about each other's people and that is what made the meeting. I really felt like I was meeting an old friend. I do not think what happened on this trip would

> have happened if we had not done that. (Mrs. Smith, Interview, March 9, 2017)

This example conveys the importance of relationship building prior to high stakes activities such as traveling to meet in person. Furthermore, it highlights the importance of building the personal and professional aspects of RLPs.

9.2 May 2017

In May 2017, Mr. Jun travelled to Toronto, Canada to spend time with Mrs. Smith at her school. Prior to making the trip, Mr. Jun had worked with Mrs. Smith and the researchers to select the topic he was going to teach in Mrs. Smith's autism spectrum program class. During planning meetings, Mr. Jun's main concern with teaching in the new environment was his lack of knowledge of the students' prior math learning, students' abilities to work in groups, and habits regarding technology and manipulatives. Mrs. Smith worked with Mr. Jun in helping him adapt the lesson so it would be suitable for her students that each had individual learning plans. In the end, Mr. Jun decided to teach a lesson on 2D geometry using tangrams as he believed that it would be accessible to all the students and fun.

Mr. Jun spent his first day in Toronto at Mrs. Smith's school, interacting with the teachers and administrators, observing math lessons in Mrs. Smith's class as well as another teacher's class and touring the school. By spending half a day immersed in Mrs. Smith's class, interacting with the students, and understanding their needs, Mr. Jun was able to adjust his personal expectations and his lesson to better suit the students' needs. He was also able to learn about the students' personalities and interests to better connect with them the following day.

During his lesson on tangrams, Mr. Jun catered to the individual needs of the students by using manipulatives, visual cues, repeating instructions visually and verbally, as well as creating flexible groupings. Reflecting on his lesson and his interaction with the class, Mr. Jun found that differentiated teaching was not as complicated as he had thought, and that these students with exceptionalities were not as different as he thought they would be (Mr. Jun, Interview, May 30, 2017). Mr. Jun had tried to incorporate some of the cross-curricular teaching strategies that he had seen Mrs. Smith use in her March lesson. In addition to teaching the students geometry content, Mr. Jun also added historical lessons and found that it "did not take away from accomplishing the goal of his lesson but resulted in students' voices being heard and a more inquiry guided way of teaching."

Mr. Jun's conversations with other teachers and researchers re-affirmed his belief in the importance of reciprocal learning. He felt that this experience motivated him to reflect deeply about his own teaching strategies and classroom culture. He said that, since Canada is a country of many immigrants from many different backgrounds, Canadian teachers are exceptional at being able to help these students to not only respect each other but also work together (Mr. Jun, Interview, May 30, 2017).

During his observations when visiting the Canadian classrooms, he consistently noted that learning in Canada seemed more flexible and freer than back home. He saw this flexibility and freedom when the students worked in different groupings, ate snacks during lessons, sat at the carpet during lessons, and when the teachers would pause their previously planned lesson to answer questions that were off topic. He compared these things to his own classroom and wondered whether these changes would have a positive effect to his students' learning.

He felt that some of the rules that he currently may not be useful for the students but more for the teacher's comfort. He hoped to change that to focus more on student needs. Furthermore, he pointed out that having a classroom setup where the students move around and have different learning areas may be beneficial for student thinking and be fairer to students who learn differently.

10 Discussion

Using a narrative approach, this study situates each participant's reciprocal learning within their classroom and schools and within their past and present contexts (Connelly et al., 1997). Each teacher's personal practical knowledge has grown throughout their 26-month experience. When considering the areas of teacher growth, we see that the teacher participants have exhibited learning in mathematics subject matter knowledge, pedagogical content knowledge, knowledge of the learner, multiculturalism knowledge, and knowledge of oneself (Ball et al., 2008; Gorski, 2009; Shulman, 1987). This newly found knowledge on mathematics education was not easily gained through the RLPs as teachers had to navigate and learn through bridging the cultural differences, interpersonal differences, and the time and space differences.

Mathematics subject matter knowledge, as outlined by Shulman (1987) and Ball et al. (2008), is found mostly in the narratives of the generalist Canadian teacher, Mrs. Smith. Mr. Jun, who is a mathematics specialist, was confident in his math subject knowledge, thus focused his learning on other areas. Through the lesson study videos, Mrs. Smith was able to gain a deeper understanding of math concepts that she taught in her own classroom. For example, Mrs. Smith shared that, while watching Mr. Jun's lesson on two-digit by three-digit multiplication, she finally understood why putting a zero in place of the ones column is so important. These finding are consistent with Ma's (1999) study that shows how subject specialist math teachers in China have higher subject matter knowledge than US teachers.

The category of PCK is vast. Researchers divide it into many separate areas such as knowledge of content and curriculum, knowledge of content and students, and knowledge of content and teaching (e.g., Ball et al., 2008; Hill, Ball, & Schilling, 2008). The majority of time when these teachers were discussing mathematics education, they were focused on sharing and learning PCK.

Examples of PCK learning in Mrs. Smith and Mr. Jun's RLP can be found throughout their asynchronous and synchronous interactions. The two teachers dedicated their time to sharing their own pedagogical approaches and to learning about and adapting their partner's strategies to their teaching. An example of this is seen when they share the ways they include hands-on crafting in their geometry lessons. The two teachers discussed how hands-on learning is beneficial for their learners in understanding geometry. They connected specific geometric principles such as Pythagoras theorem and 3D shapes to these teaching tools. Furthermore, by using geometry-focused crafts, they are able to motivate their learners to engage in mathematics without even realizing it. This learning would be categorized by Ball et al. (2008) not only as knowledge of content and learning but also as knowledge of content and teaching.

Furthermore, the teachers also took time to discuss knowledge pertaining to content and curriculum (Ball et al., 2008). The teachers shared with each other the key elements of their elementary mathematics curriculum. After taking the time to compare and contrast structural similarities and differences, the teachers also shared their experience and opinions on the curriculum they use.

When learning about each other's curriculum, the teachers talked about ways in which their partner's curriculum may be beneficial for mathematics teaching and learning and made decisions to modify their teaching in those ways while still staying within the bounds of their own curriculum. An example of this is exhibited when Mrs. Smith learned about the introduction of fractions, ratios, and percentages in Chinese curriculum. She feels that introducing them one at a time, as outlined by Chinese curriculum documents and textbooks, is more advantageous to students as these difficult topics are easily confused. She believes that, when she teaches these topics in the future, she will dedicate more time to them and take a more one by one approach as opposed to comparing and contrasting fractions, ratios, and percentages from the beginning.

Learning from each other's curriculum is also apparent when teachers discuss the importance of "foundational knowledge." There was general agreement that more time should be spent on foundational knowledge and basic computation skills early on in elementary school. Once students have that foundational knowledge, they can then diversify their learning. This focus on foundational skills is more evident in the Chinese curriculum (Ministry of Education of the People's Republic of China, 2012) as it divides elementary math in two stages, which focus heavily on numbers and algebra and spatial reasoning.

The most impactful RLP activity for increasing teacher knowledge of the learner was the teacher exchange that occurred between Mrs. Smith and Mr. Jun. More specifically, the teachers' experiences teaching in each other's classroom led to new understanding of learners. When Mr. Jun was challenged to teach in a classroom comprised entirely of students with autism, he was initially very worried that he was not equipped with the appropriate pedagogies. However, after learning about autism spectrum disorder and the individual learning plans of Mrs. Smith's students, Mr. Jun tackled the teaching opportunity with a hands-on and visual tangrams lesson. When reflecting on his experience, Mr. Jun states "these students are not that special" by which he means that, although they have their individual needs, he felt as though they are still children, and many of the pedagogies he uses for these students can be used to teach them.

When Mrs. Smith taught in Changchun, she observed the level of comfort the Chinese learners had with visitors in their classroom and the readiness they exhibited when asking for help, from their teacher and the visitors. Mr. Jun suggested that this confidence and enthusiasm comes from the students feeling like they are the center of the school.

Furthermore, because the students have multiple teachers, they feel that all teachers are their teacher.

Teacher efficacy increased when teachers learned that the reciprocal learning partner was utilizing the PCK that they had shared. An example of this is seen when Mrs. Smith felt validated when Mr. Jun used her circumference lesson resources. To her, having an exemplary teacher use her resources was a confirmation that she too is a good math teacher.

When the teachers were taken out of their comfort zones and were challenged with teaching tasks that they had previously felt were beyond their ability, their success increased their self-efficacy. An example of this is seen during the teacher exchange activities when Mrs. Smith and Mr. Jun were challenged to teach in learning environment that they would previously have found challenging. When Mr. Jun succeeded in teaching an effective lesson in Mrs. Smith's special education class, his teacher efficacy regarding teaching learners with exceptionalities increased.

11 Conclusion

The reciprocal learning partnership gave Mr. Jun and Mrs. Smith opportunities to learn new mathematics pedagogies that were cross-cultural and beneficial to their classes. Additionally, both teachers took their learning beyond mathematics teaching and also discussed topics on classroom management, student grouping, differentiated teaching, and special education. Mrs. Smith felt that watching Mr. Jun's lessons also improved her knowledge of specific math content areas.

In addition to learning from each other through synchronous communications such as video conferencing, the two teachers communicated effectively through asynchronous modes such as WeChat and lesson studies. These modes of communication were effective because they were more visual and this cut across language barriers and miscommunication due to translations. Asynchronous student interactions between the teachers' classes were initially a big motivation for the teachers and, although they were not consistently sustained throughout the RLP, they functioned as a way for the teachers to bring their learning back to the classroom in a way that was exciting for the students.

For both teachers, developing a relationship with each other before visiting and teaching in the other's class was essential in developing a sense of comfort. Building the relationship meant that the teachers communicated on topics that were not isolated to mathematics education and teaching. By sharing personal information about their families, the teachers formed a stronger bond that helped sustain their RLP. While language barriers and school schedules often got in the way of communication, both teachers remained in their RLP because they felt a sense of responsibility toward each other and as representatives of their school and country.

References

An, S. (2008). Outsiders' views on Chinese mathematics education: A case study on the United States teachers' teaching experience in China. *Journal of Mathematics Education, 1*(1), 1–27.

Anderson-Levitt, K. (2008). Globalization and curriculum. In F. M. Connelly (Ed.) with M. F. He and J. Phillion (Assoc. Eds.), *The Sage Handbook of Curriculum and Instruction* (pp. 349–368). Thousand Oaks, CA: Sage Publications Inc.

Atweh, B., & Clarkson, P. (2001). Internationalisation and globalization of mathematics education: Towards an agenda for research/action. In B. Atweh, H. Forgasz, & B. Nebros (Eds.), *Sociocultural research on mathematics education: An international perspective* (pp. 77–94). New York: Erlbaum.

Ball, D. L., Thames, M. H., & Phelps, G. (2008). Content knowledge for teaching: What makes it special? *Journal of Teacher Education, 59*, 389–407.

Beech, J. (2009). Who is strolling through the global garden? International agencies and educational transfer. In R. Cowen & A. M. Kazamis (Eds.), *International handbook of comparative education* (pp. 341–357). The Netherlands: Springer.

Brickman, W. (1960). A historical introduction to comparative education. *Comparative Education Review, 2*(3), 6–13.

Cai, J., Ding, M., & Wang, T. (2013). How do exemplary Chinese and U.S. mathematics teachers view instructional coherence? *Educational Studies in Mathematics, 85*(2), 265–280.

Clandinin, D. J. (Ed.). (2007). *Handbook of narrative Inquiry: Mapping a methodology*. Thousand Oaks, CA: Sage Publications.

Connelly, F. M., & Clandinin, D. J. (1990). Stories of experience and narrative inquiry. *Educational Researcher, 19*(5), 2–14.

Connelly, F. M., & Clandinin, D. J. (2006). Narrative inquiry. In J. Green, G. Camilli, & P. Elmore (Eds.), *Handbook of complementary methods in education research* (pp. 375–385). Mahwah, NJ: Lawrence Erlbaum.

Connelly, F. M., Clandinin, D. J., & He, M. F. (1997). Teachers personal practical knowledge on the professional knowledge landscape. *Teaching and Teacher Education, 13*(7), 665–674.

Connelly, F. M., & Xu, S. (2015). Reciprocal learning: Comparative models and the partnership project. In *Reciprocal learning & symbiotic development* (pp. 1–18). Shanghai.

Gorski, P. (2009). What we're teaching teachers: An analysis of multicultural teacher education coursework syllabi. *Teaching and Teacher Education, 25*(2), 309–318.

Haas, B., & Weale, S. (2017). Chinese maths textbooks to be translated for UK Schools. *The Guardian*. Retrieved from https://www.theguardian.com/education/2017/mar/20/chinese-maths-textbooks-to-be-translated-for-uk-schools.

Hall, D. L., & Ames, R. T. (1999). *Democracy of the dead*. Chicago: Open Court Press.

Hayhoe, R. (2008). Philosophy and comparative education: What can we learn from East Asia? *Comparative and international education: Issues for teachers.* Toronto & New York: CSPI & Teachers College Press.

Haythornthwaite, C. (2002). Building social networks via computer networks: Creating and sustaining distributed learning communities. In K. A. Renninger & W. Schumar (Eds.), *Building virtual communities: Learning and change in cyberspace*. Cambridge: Cambridge University Press.

Hill, H. C., Ball, D. L., & Schilling, S. G. (2008). Unpacking "pedagogical content knowledge": Conceptualizing and measuring teachers' topic-specific knowledge of students. *Journal for Research in Mathematics Education, 39*, 372–400.

Huang, J. (2017). *A narrative inquiry into Chinese pre-service teacher education and induction in Southwest China through cross-cultural teacher development*. Retrieved from http://scholar.uwindsor.ca/cgi/viewcontent.cgi?article=6943&context=etd.

Karp, A. (2013). From the local to the international in mathematics education. In M. A. Clements, A. J. Bishop, C. Keitel, J. Kilpatrick & F. K. S. Leung (Eds.), *Third international handbook of mathematics education* (pp. 797–826). New York: Springer Science + Business Media.

Kilpatrick, J. (2013). Part IV introduction to Section D: International perspectives on mathematics education. In M. A. Clements, A. J. Bishop, C. Keitel, J. Kilpatrick, & F. K. S. Leung (Eds.), *Third international handbook of mathematics education* (pp. 791–795). New York: Springer Science + Business Media.

Ma, L. (1999). *Knowing and teaching elementary mathematics: Teachers' understanding of fundamental mathematics in China and the United States*. New Jersey: Lawrence Erlbaum Associates Inc.

Merriam, S. B. (1998). *Qualitative research and case study applications in education*. San Francisco, CA: Jossey-Bass.
Miller, K. F., Kelly, M. K., & Zhou, X. (2005). Learning mathematics in China and the United States Cross-cultural insights into the nature and course of mathematical development. In J. I. D. Campbell (Ed.), *Handbook of mathematical cognition* (pp. 163–178). New York: Psychology Press.
Ministry of Education of the People's Republic of China. (2012). *Mathematics curriculum standards for compulsory education (2011 Edition: 4–5)*. Beijing: Beijing Normal University Press.
Ram, A. (2006). *A multilevel analysis of mathematics literacy in Canada and Japan: The effects of sex differences, teacher support, and the schooling learning environment.*. Retrieved from http://dspace.library.uvic.ca:8080/handle/1828/986.
Sadler, M. (1979). How far can we learn anything of practical value from the study of foreign systems of education? Address given at the Guildford educational conference on Saturday 20 October 1900. In J. H. Higginson (Ed.), *Selections from Michael Sadler: Studies in world citizenship* (pp. 48–51). Liverpool: Dejall & Meyorre.
Schubring, G. (2003). L'Enseignement Mathématique and the First International Commission (IMUK): The emergence of international communication and cooperation. In D. Coray, F. Furinghetti, H. Gispert, B. Hodgson, & G. Schubring (Eds.), *One hundred years of L'Enseignement Mathématique: Moments of mathematics education in the twentieth century* (pp. 47–66). Geneva, Switzerland: L'Enseignement Mathématique.
Shulman, L. S. (1987). Knowledge and teaching foundations of the new reform. *Havard Educational Review, 57*(1), 1–22.
Singh, P., & Ellerton, N. F. (2013). International collaborative studies in mathematics education. In M. A. Clements, A. J. Bishop, C. Keitel, J. Kilpatrick, & F. K. S. Leung (Eds.), *Third international handbook of mathematics education* (pp. 827–860). New York: Springer Science + Business Media.
Stake, R. E. (1995). *The art of case study research*. Thousand Oaks, CA: Sage.
Xu, S. J. (2011). Bridging the East and West dichotomy: Harmonizing Eastern learning with Western knowledge. In J. Ryan (Ed.), *Understanding China's education reform: Creating cross cultural knowledge, pedagogies and dialogue* (pp. 224–242). London, UK: Routledge.
Xu, B. (2016). *Math stories in elementary mathematics education in China and North America*. Retrieved from https://ir.lib.uwo.ca/cgi/viewcontent.cgi?article=5611&context=etd.
Xu, S. J., Chen, S., & Huang, J. (2015). Pedagogies of diversity: West-East reciprocal learning in pre-service teacher education. In L. Orland-Barak & C. J. Craig (Eds.), *International teacher education: Promising pedagogies* (Part

B: Advances in research on teaching) (Vol. 22B, pp. 137–160). Bingley, UK: Emerald Group.

Xu, S. J., & Connelly, M. (2010). Narrative inquiry for school-based research. *Narrative Inquiry, 20*(2), 349–370.

Xu, S. J., & Connelly F. M. (Project Directors) (2013). *Reciprocal learning in teacher education and school education between Canada and China.* Social Sciences and Humanities Research Council of Canada (SSHRC) 2013–2020 [*Grant 895-2012-1011*].

Yazan, B. (2015). Three approaches to case study methods in education: Yin, Merriam and Stake. *The Qualitative Report, 20*(2), 134–152.

Zhu, Z., & Monroe, E. E. (1991). Mathematics education in China today: Four problem areas. *Educational Studies in Mathematics, 22*(2), 205–208.

CHAPTER 6

Professional Development in Mathematics Education Through Reciprocal Learning: The Case of Mrs. Yang and Ms. Ko

Sijia Cynthia Zhu

1 Introduction

This chapter tells the story of Mrs. Yang and Ms. Ko, two elementary teachers from Changchun, China and Toronto, Canada, who spent three years developing a reciprocal learning partnership (RLP) for the purpose of professional development in mathematics teaching and learning. RLP is a form of professional development in which teachers partner with each other with the shared goal of learning from and with one another. The partnership creates a space for teachers to connect directly to explore each other's knowledge, values, and teaching methods (Xu & Connelly, 2013). In the case of Mrs. Yang and Ms. Ko, this relationship was a cross-cultural learning experience focused on elementary mathematics.

S. C. Zhu (✉)
Ontario Institute for Studies in Education, University of Toronto, Toronto, ON, Canada
e-mail: cyn.zhu@mail.utoronto.ca

S. C. Zhu et al. (eds.), *Reciprocal Learning for Cross-Cultural Mathematics Education*, Intercultural Reciprocal Learning in Chinese and Western Education,
https://doi.org/10.1007/978-3-030-56838-2_6

When examining how the teacher engages in reciprocal learning, we can turn to past research on types of teacher knowledge to consider what they may learn during their experiences. Teacher knowledge was first introduced by Shulman (1987) as content knowledge, general pedagogical knowledge, curriculum knowledge, pedagogical content knowledge (PCK), knowledge of learners and their characteristics, knowledge of educational contexts, knowledge of educational ends, purposes, and values, and their philosophical and historical grounds. Since Shulman first developed a model for teacher knowledge, many other models have emerged such as Ball, Thames, and Phelps' (2008) mathematics knowledge for teaching (MKT) model, Connelly, Clandinin, and He's (1997) personal practical knowledge, as well as Holden and Hicks' (2007) and Gorski's (2009) teacher knowledge of global education and multiculturalism.

For mathematics, Ball et al.'s (2008) MKT model divides teacher knowledge into subject matter knowledge and PCK. Subject matter knowledge is then divided into common content knowledge, which is the knowledge of mathematical content; specialized content knowledge, which is mathematical knowledge and skills unique to teaching such as being able to identify patterns in student errors; and horizon content knowledge, which is awareness of the relationship between math topics (Ball et al., 2008). Ball et al.'s (2008) PCK is divided into knowledge of content and students, content and curriculum, and content and teaching.

Connelly and Clandinin (1988) believe that personal practical knowledge is founded on a "teacher's past experience, in the teacher's present mind and body, and in the future plans and actions" (p. 25). According to Ben-Peretz (2011), this expands upon the common subject matter or pedagogical knowledge perspective to a more unique type of teacher knowledge that is situated in narrative teacher stories. Connelly et al. (1997) state that "a rich, deeper, more narrative understanding evolves from studying what we term the professional knowledge landscape. To understand teaching, we need to understand it in a complex environment" (p. 273).

As we enter the twenty-first century, an age of increased globalization, the amount of teacher knowledge that is necessary in order to help students make sense of the world is greatly expanding (Holden & Hicks, 2007). Furthermore, Tye (1999) suggests that education, in this globalized time, involves learning about societal issues that transcend national boundaries and that are cross-cultural. In order for teachers to prepare students, they must have knowledge of strategies for teaching about the global and controversial issues, and knowledge for evaluating sources of information (Holden & Hicks, 2007).

In addition, Gorski (2009) suggests that an important societal issue in this age of globalization is teaching to a multicultural context, which requires a specific teacher knowledge. This multicultural teacher knowledge includes: "knowledge for (1) Teaching the 'other', (2) Teaching with tolerance and cultural sensitivity, and (3) teaching with multicultural competence" (Gorski, 2009, p. 312).

One way in which teachers gain teacher knowledge once they are in-service teachers is through professional development. When evaluating the effectiveness of professional development opportunities, it has been found that one-time sessions and lectures on best practices are largely unsuccessful in achieving sustained improvement and motivation among teachers (Arbaugh, 2003; Gojmerac & Cherubini, 2012). According to research on professional development, successful interventions require a combination of job-embedded learning, collaborative inquiry, institutional and administrative support, and of course teacher commitment to continuous, sustained, and intensive engagement in professional development initiatives (Darling-Hammond, Hyler, Gardner, & Espinoza, 2017; Desimone, 2011; Guskey & Huberman, 1995).

Furthermore, Schlager and Fusco (2004) suggest that the foundation of successful professional development programs is the development of a community. Wilson and Berne (1999) also concluded that effective efforts for professional development "all (a) involve communities of learners redefining teaching practices, (b) seek to activate (rather than deliver) teacher learning, and (c) privilege teachers' interactions with one another" (p. 194). Long-term RLPs between Canadian and Chinese mathematics teachers fulfill all three criteria that Wilson and Berne (1999) outline as necessary for effective professional development (Connelly & Xu, 2015).

Collaborative teacher inquiry and communities of practice are methods in which professional development can privilege teacher interactions with one another. Nelson and Slavit's (2008) framework for collaborative teacher inquiry suggests that successful endeavors allow for a cyclical process done by teachers in a school context. This cyclical process allows for teachers to focus their professional development on issues that occur in their teaching environment and develop action plans that are carried out, analyzed, reflected upon, and adapted while in dialogue with their peers (Nelson & Slavit, 2008).

When successful, collaboration can lead to effective and collective implementation of instructional innovations as well as assisting in making curricular change (Achinstein, 2002), which can benefit student performance, teacher professional performance, and school culture (Jao & McDougall, 2016). However, setting up effective collaborative teacher inquiries require motivated and hardworking teachers and a support system and collaborative structure that can overcome interpersonal, temporal, and logistical barriers (Jao & McDougall, 2016). Jao and McDougall (2016) suggest that it is important for collaborative teacher inquiry teams to create team goals and for administrators to schedule regular "collaborative time." Furthermore, using an online platform for collaboration can "create opportunity for the collaborations that may be logistically difficult" (Jao & McDougall, 2016, p. 568).

2 Methods

This research stems from Xu and Connelly's (2013) Social Science and Humanities Research Council-funded partnership project, Reciprocal Learning in Teacher Education and School Education (2013–2020) that focuses on the development of cross-cultural knowledge and relationships in all areas of primary and secondary education, with the goal of improving student achievement in both nations. Researchers from four Chinese universities and two Canadian universities joined together with elementary and secondary school teachers, and administrators to implement long-term RLPs between Canada and China.

The RLP is comprised of one elementary mathematics teacher from Changchun, one elementary mathematics teacher from Toronto, and a facilitator and translator. While the RLP described in this chapter involves student interactions, they are only considered to be part of the RLP as an extension of their teacher's participation.

Given the long-term and exploratory nature of RLPs, it was crucial to find elementary mathematics teachers who were willing to commit to the RLP for multiple school years and who were comfortable working with researchers. As this study focuses on mathematics education, participants who demonstrated interest in seeking professional development in mathematics were sought out. Furthermore, all of the participants were selected due to recommendations by their administrators or by researchers for their efforts in mathematics teaching. Lastly, due to the geographic separation between Toronto and Changchun, it was essential for the teachers to be comfortable and willing to use e-communications as a primary means of communication.

Ideal participants were found in Toronto by reaching out to teachers that had participated in multiple mathematics focused professional development opportunities at the University of Toronto. Participants in Changchun were selected from teachers that were recommended by administrators at the Primary School attached to NENU (PNENU). In this chapter, I present the case studies of Ms. Ko and Mrs. Yang.

This project is grounded in a narrative framework that focuses on each teacher's personal experiences as well as cultural and historical contexts when participating in RLP with a mathematics education focus (Andrews, 2007; Connelly & Clandinin, 1990; Creswell, 2012; Fox, 2006). Researchers collaborated with the participants in planning the program of data collection to assist in telling their story (Creswell, 2012). Data was collected over a three-year period, from December 2014 to June 2017.

From December 2015 to March 2015, data was collected from participants with the hopes of matching them to a RLP and mainly consists of introductory interviews, teaching videos, and school information. Starting in March 2015, data was collected from the RLP consisting of interviews, teaching videos, e-communications between the participants, journals, and notes on observations, and conversations (Stake, 1995; Yazan, 2015).

As this was an exploratory study done through narrative inquiry, the data was collected as the teacher pairs developed their relationships and communicated on their own timeline. Therefore, documents, digital communication, and interviews were not collected at set times but rather as the opportunities naturally occur.

Following the narrative framework, all of the data collected from each reciprocal learning case were looked at together as a whole in order to retell the story of each reciprocal learning teacher pair (Stake, 1995). The stories focus on each teacher pair's lived experiences and are told in chronological order. The historical, cultural, and personal contexts of the teacher pairs created the setting for the stories (Connelly & Clandinin, 2006; Creswell, 2012). In keeping with the narrative approach, the stories collected from the teacher pairs and the themes that emerge are the source of the major findings of the study (Creswell, 2012).

3 Getting to Know Ms. Ko

Ms. Ko attended the University of Toronto in Ontario, Canada, and completed a bachelor in science degree majoring in psychology and minoring in sociology and history. Initially, Ms. Ko had aspired to become a psychologist; however, as she describes it, she "fell into teaching serendipitously" while working at the YMCA and discovering that she loved children and teaching. Once she settled on teaching as her career goal, Ms. Ko went back to the University of Toronto and completed a bachelor of education degree in the junior and intermediate division, specializing in history. Since graduating from her bachelor of education degree in 2000, Ms. Ko has taught at three different schools, one private and two public schools. She has been teaching at her Harbord Street School (HSS) for eight years. Over her 15-year teaching career, she has taught English, history, mathematics, social studies, science, art, dance, music, physical education, and English as a second language. She has been teaching mathematics for 14 years. During her time at HSS, Ms. Ko went to graduate school and earned a master's degree in education. During the course of this project, she completed a Ph.D. degree in mathematics education.

During the span of the project, Ms. Ko taught grade five from 2014 to 2015 and again in 2015–2016. During the 2016–2017 school year, Ms. Ko taught a grade four and five split class. In all of these classes, she was responsible for teaching mathematics and science among many other subjects. Through all three years, her classes have been comprised of students coming from many ethnic and cultural backgrounds. Ms. Ko describes the community that she serves as one comprised of many immigrant families who care deeply about their children's learning.

From my observations in her class, Ms. Ko is extremely respected by her students and the staff at her school. Her classroom management is particularly notable as, year after year, her class and her students are always attentive, happy, and caring toward each other. I mentioned this observation to Ms. Ko, and she shared that this was not always the case and that, when she first began her teaching career, she had trouble with classroom management.

Those troubles are why she started developing strategies that drew upon her background in psychology. She shares that it took her a few years, but eventually, she was able to develop a set of strategies and routines that ensured her students a safe and happy learning environment. In her own words, she describes that the most rewarding part of teaching is "seeing students get excited about learning and witnessing their self-esteem blossom. Making personal connections with students always brings me great joy." However, as class sizes increase, she finds that the large numbers make it difficult for her to meet the needs of each student, especially those with exceptionalities.

While Ms. Ko has spent a great deal of her time focusing on mathematics education both in her own teaching and in her Ph.D. degree, she does not think of herself as someone who is good at math. In fact, she commented on multiple occasions that she is an Asian who is bad at math. She described that, as a child, she often felt that she was bad at math and experienced self-doubt and embarrassment for her poor math skills because she associated being good at math with being intelligent.

As a teacher, she did not want any of her students fearing and dreading math; thus, mathematics and the teaching of it became a personal and passionate subject for her. When it comes to elementary mathematics, she finds that her students often have difficulties with metric conversions. She finds that it is helpful to teach students mnemonics to help them remember the order; however, she feels that it does not help them gain a deeper understanding for the principles behind making conversions.

In addition to teaching her own core class, Ms. Ko is also very involved in the school. During recesses, lunches, and after school, she runs the Sign Language Club, Future ACES, and the girls' soccer team. Recently, she has also started the African/Caribbean Leadership Club (ACLC) to help some boys at the school that have been disruptive in class. In addition to running extracurricular activities for the students, Ms. Ko is also part of the school's School Improvement Plan Committee and the Mental Health and Wellness Committee.

As Ms. Ko goes about her work at HSS, she feels that her relationship with her coworkers is what leads her to feel so supported. On multiple occasions, she has described her coworkers to be like her family. Very often, she spent her lunch hour breaks meeting with her fellow teachers to discuss school-related issues, concerns, and ideas. In addition to working with her fellow teachers, Ms. Ko has also participated in professional development opportunities within the school such as working with learning coaches and specifically math coaches on topics such as three-part lesson plans and number talks.

Ms. Ko was teaching full time and heading many extracurricular groups at her school and also completing her Ph.D. degree in mathematics education during the time span of this project. That combination made her very busy, and often she apologized for not being able to make time for this project. Yet, she continued to work with Mrs. Yang for three school years and consistently put in the effort to involve her students in the project.

4 Getting to Know Mrs. Yang

Mrs. Yang went to university in Changchun, China, and attended Northeast Normal University (NENU), graduating with a bachelor's degree in elementary education in 2006. When Mrs. Yang had first applied to university, she had not planned on going to a normal university or becoming an elementary teacher. However, when her original plans did not work out, she ended up attending NENU. While she was at NENU, she found that she began to really enjoy working with students and became more and more interested in being a teacher.

After graduating from university, Mrs. Yang earned a job at the Primary School attached to NENU (PNENU) teaching elementary mathematics. At the beginning of this project, Mrs. Yang had been teaching at PNENU for seven years and had taught mathematics to students in grades one to four. During the first year of the project, Mrs. Yang was teaching two classes of grade four mathematics. She was part of a five-person teaching team that consisted of mathematics teachers teaching to other classes of grade four students.

One of the teachers on her teaching team, Mr. J, is also part of this project. It is common for teachers to advance with their students from year to year, so, for the latter two years of the project, Mrs. Yang taught the same two classes of students as they went from grade four to five and eventually to grade six and graduation. Currently, she has returned to teaching grade four mathematics.

Throughout her time at PNENU, Mrs. Yang has eagerly participated in many different types of professional development (PD) opportunities. Mrs. Yang describes that some of these opportunities are mandatory for all teachers while some are elective or through school administration recommendation. Participation in provincial and city-wide PD sessions is usually through an online portal, and the topics are focused on either education and teaching or a subject-specific to mathematics. In addition to online

PD, Mrs. Yang shares that her school also has their own PD opportunities such as in-class model teaching and teacher as researcher seminars.

When she first began her career as a teacher, Mrs. Yang shares that the most challenging part of her work was relating to the students. At that time, she did not have her own children and her experience working with young students was minimal, so she found that there was a barrier in communication. Mrs. Yang found that the students could not understand her when she used her teacher language, so she sought out opportunities where she could learn the children's language.

In terms of mathematics, Mrs. Yang finds that the division of decimals, fractions, and ratios to be the most difficult topics in elementary mathematics for students to understand. Consequently, this means that those topics are the most difficult for her to teach. In order to help her students understand mathematics, she worked with other teachers on her teaching team to enhance and rearrange specific lessons. She believes that, in order for students to understand these topics, they have a firm grasp on meaning behind the operations.

In addition to the individual preparation she does for her teaching, Mrs. Yang also works with five other teachers that teach the same grade. Every week, they meet on Tuesday afternoons for approximately 90 minutes to prepare for upcoming lessons. She finds group preparation to be greatly beneficial because they are able to share their resources, experiences, and perspectives. However, it also has drawbacks as it sometimes leads to passiveness, and she does not think deeply about problems herself.

Furthermore, while teaching at PNENU, she enrolled in NENU to complete a master's degree in Theory of Education. She often mentioned books that she was reading, not just for mathematics education, but also about children's development and psychology. Multiple times throughout this project, Mrs. Yang was selected by her administrators as the teacher representative to participate in professional development opportunities and model teaching seminars. Mrs. Yang has also created teaching videos that are accessible to all teachers online as a model for mathematics teaching.

In addition to teaching mathematics, Mrs. Yang is also responsible for leading a special interest group for students. The group that she leads is the dragon craft group where students use different crafting techniques to create three-dimensional figures of dragons.

5 Learning About, from, and with Each Other

Mrs. Yang and Ms. Ko's professional development through RLP started in January 2015. They first began by getting to know each other on a personal level through a series of surveys that they each filled out and shared with their partners. They then went on to expand their knowledge of one another's teaching contexts through sharing documents such as curricula, lessons plans, and textbooks. The two teachers would meet through synchronous video chat or text chat and explore their newly acquired information and ask each other follow-up questions. While the two teachers learned a great deal through these interactions, there were two other RLP activities that most benefited their professional development, learning through asynchronous lesson study and including their students in their RLP.

5.1 Learning Through Asynchronous Lesson Study

During interviews with Mrs. Yang and Ms. Ko, both teachers had revealed that they hoped to see each other's teaching in action. Therefore, in 2015 and 2016, both teachers participated in a lesson study process of their RLP's teaching. The interview questions and content topics for the lesson study were not chosen by the teachers, but were selected through consideration of their teaching interests. Mrs. Yang watched a series of videos of Ms. Ko teaching the topic of three-digit by two-digit multiplication, while Ms. Ko watched Mrs. Yang teaching a lesson on equivalent relationships.

Both teachers were very complimentary toward each other regarding the lesson they watched. Ms. Ko says:

> To me, it speaks to her teaching in general where she was able to prep her kids to the point where, in this type of context, they are doing the inquiry-based. It was like clockwork. They all knew what to do, and it flowed perfectly. So amazing. That was like a snapshot of what inquiry-based learning should look like. (interview, August 4, 2016)

Mrs. Yang spoke similarly, sharing that she thought Ms. Ko taught her lesson very well. She noticed that Ms. Ko was able to build on the foundation of each student's independent thinking and, in very detailed ways, teach them to calculate multi-digit multiplication problems. She also noticed Ms. Ko was able to teach while displaying the learning visually and that she took every opportunity to ask questions with her students (interview, April 21, 2016). These positive comments indicate that both teachers found value in each other as teaching professionals.

In addition to positive feedback, both teachers reflected upon their own teaching. After watching the way Mrs. Yang integrated inquiry-based learning and communicated with her students, Ms. Ko said:

> I think this is where I fail, to trust the kids. I always feel that I need to give it to them and they let them practice what I give them, but it is actually elicited from them. And let them teach each other. So I always assume that when the kids teach each other, I have to repeat everything. She does that, but they have the flow and she just clarifies, like, ok, they have shared and now I have to make sure that they hear it correctly. (interview, August 4, 2016)

Mrs. Yang also reflected on her own teaching saying that, after watching Ms. Ko's students sitting on the carpet focused on the lessons and working so productively in free groups, she wonders if her own classes are too structured for her students (interview, April 21, 2016).

In addition to the differences in their teaching, both teachers picked up on similarities that they believed made them successful. They both commented that was interesting to see that, regardless of culture, independent thinking and focusing on students' different personalities and needs were important aspects of both of their teaching. Furthermore, Mrs. Yang stated that, if she were to teach the same lesson (two-digit by three-digit multiplication), she would have focused on the same key learning points (interview, April 21, 2016).

The two teachers both focused mainly on pedagogy when observing the lessons. Ms. Ko made only one comment about the math content, remarking that Mrs. Yang did a good job in "conveying that there are multiple strategies of arriving at the same answer, but as well the parentheses was critical" (interview, August 4, 2016). Mrs. Yang questioned why Ms. Ko allowed her students to use repetitive addition as a method for achieving their answer in multi-digit multiplication as she believed that the method has a higher chance of error and becomes much too complicated as the numbers get bigger (interview, April 21, 2016).

While the videos gave the teacher a clear snapshot of how their partner teaches and a specific topic lesson to learn from, the static video was not able to answer all of the teachers' questions. Both teachers' interview answers were riddled with follow-up questions such as "how does Mrs. Yang set her students to work so well independently and in groups? (interview, August 4, 2016)" and Mrs. Yang asking about how Ms. Ko decides on her lesson format when she has so much flexibility in her teaching schedule. Overall, Mrs. Yang's questions were much more critical of Ms. Ko's teaching, while Ms. Ko's questions were more exploratory in nature. This plethora of questions after watching the video indicated that it would be useful for the teachers to have time to discuss these lessons with each other after watching them.

6 Learning with and Through Their Students

Starting in April 2015, Mrs. Yang and Ms. Ko's students started writing letters to each other. Their communication went back and forth until December 2016. The idea of having students communicate was first brought up in Ms. Ko's class when she told the students about her participation in the RLP. Her students were very excited about her participation and very excited about having a researcher in their classroom. They asked Ms. Ko if she would go to China and if they would have the opportunity to meet the Chinese students.

As a way of giving the students an opportunity to interact with the project, Ms. Ko and I came up with the idea of pen pals and sending a class mascot to experience school in China. Mrs. Yang agreed to the exchange activities during our April 1, 2015, Skype meeting, and thus, the interactions between students started. In this section, I will focus on student interactions and how they play an essential role in RLP and teacher development and learning.

From April 1, 2015, to December 2016, Ms. Ko and Mrs. Yang's students wrote letters, math problems, and sent artifacts back and forth to each other a total of six times each. In most cases, the letters were mailed as a package from the teachers, but, in some cases, the letters were delivered by researchers travelling to the cities. At different times, the teachers gave their students different prompts for the letters and responses largely depended on the letters the students had previously received. Students were mostly given freedom to write on the topics that they were interested in; however, the teacher strongly encouraged their students to communicate their reflections on mathematics learning.

English was the primary language for student communication as the Chinese students could read and write English while the Canadian students did not understand any Mandarin language. However, due to language barriers, communications were sometimes difficult between the children. The Chinese students had been studying English since grade one and could read the English letters but found it difficult to articulate themselves fully in English past the introductory stage. In those cases, letters were often written in Chinese and then translated with the help of parents or the English teacher.

During the period where Mrs. Yang and Ms. Ko's students interacted, Ms. Ko taught three different classes of students while Mrs. Yang continued to teach the same class as they advanced from grade four to grade six. This meant that every year, Mrs. Yang's class had to reintroduce themselves to a new class of students resulting in some repetitive communications. This combined with the extra efforts in writing in English and then translating into English was sometimes demotivating for the students and difficult for not only Mrs. Yang but also the students' homeroom and English teacher.

Even though it was sometimes difficult to complete the letters, Mrs. Yang class remained in communication with Ms. Ko's class. Mrs. Yang and the homeroom teachers that she collaborated with felt that the struggle was useful for the students' growth and the practice writing in English was also helpful. Furthermore, as teachers, they felt that they were responsible for opening a window to the outside for their students. Mrs. Yang also felt that it was good practice for her students to challenge their ideas about mathematics by articulating their thoughts in English as long as it did not take away from planned mathematics lessons.

Another reason that Mrs. Yang continued with the activity was that reading the letters from Ms. Ko students was useful for her learning. Furthermore, every time the letters arrived in her class, it was very exciting for her and her students, causing her a boost in motivation for staying connected with Ms. Ko. This boost in motivation was evident in that Mrs. Yang would often reach out for updates either before or after a set of letters had been mailed or delivered. Ms. Ko states:

> My students absolutely LOVED the letters and the panda, friendship!!!! After Doug presented the letters and the bear, things at school became quite busy with trips, school performances among other things, so the students did not get a chance to share the letters they received with the

class. My students wanted to read them all out loud to each other. As things were so busy, I told them that we would share the letters later. The students bothered me EVERY DAY to share their letters. We students share on different days. I could tell that they felt so connected to their pen pals!! The panda was so cherished and loved by all of my students! (online-conference, July 2, 2015)

It is evident that Ms. Ko was excited about her students' love for the exchange. Her students' motivation for the exchange continued throughout her three-year involvement in the RLP. Friendship, the stuffed animal mascot that was exchanged, become a consistent fixture in her classroom.

In many aspects of the RLP, Ms. Ko deferred to Mrs. Yang and the researchers to make suggestions for activities or discussion topics. However, when it came to student activities, Ms. Ko had many different suggestions. She consistently suggested new ideas that would stimulate the students on both sides to articulate their feeling about mathematics, to learn more about each other, and to challenge each other's mathematics problem-solving abilities. She also recommended her students to use mathematics language to communicate with their pen pal partners.

Ms. Ko's suggestions led to both classes expanding beyond writing letters to each other. In April 2015, Ms. Ko class sent Mrs. Yang class a series of math story problems, and, in April 2016, they sent a class poster that shared factoids about Toronto and Canada that were communicated using mathematical language. Mrs. Yang's class also responded with sharing an integrated math, science, and English project in April 2016 as well as by writing questions regarding mathematics learning in December 2016.

The communications between the teachers' classes also gave Ms. Ko an authentic insight into education in China and Chinese culture. Seeing the students' work on the letters, Ms. Ko got a more informal glimpse into students' abilities. Ms. Ko describes that, through reading the students' letters, she realized that Ms. Yang's students were also very diverse learners and at many different levels, unlike her previous belief that Chinese students were all excellent and high achieving. Her realization of this diversity among Chinese students translated to discussion questions in her Skype conversation with Ms. Yang.

Lastly, through prompting and encouragement by Ms. Ko, Mrs. Yang planned an activity for her students to explore the integration of science, math, and language that they could share with their sister class in Canada. The integration activity was difficult to organize, and at times frustrating for Mrs. Yang as she felt there was no time in the schedule to accommodate such activity. However, as she guided her students through the activity, she commented on how it was a good opportunity for students and for teachers to collaborate.

This activity gave Mrs. Yang a chance to put into action some of the integrated teaching techniques that she had seen in Ms. Ko's teaching that she had commented on being interesting but not possible in the Chinese setting. Upon completion of the activity, Mrs. Yang stated that it was a good activity and very good for students to experience the integration of math and English in science. She felt that she could have taught the activity better in the future as she gained experience with this type of teaching.

7 Discussion

Inquiry-based learning for mathematics is an area that Mrs. Yang and Ms. Ko explored extensively once they discovered their intersecting interest. Ms. Ko found Mrs. Yang's use of one rich task to be very effective and also used this strategy in her own classroom. By allowing students to focus on a multiple-step, inquiry-based problem and develop a routine, both teachers found learning to be more effective.

Mrs. Yang also learned about the importance of differentiation for two-digit by three-digit multiplication while watching Ms. Ko teach her lesson, as Ms. Ko had chosen to pose three-digit by three-digit and three-digit by five-digit questions later in her lesson. Mrs. Yang was inspired to challenge her students after seeing Ms. Ko's strong students tackle these problems with vigor.

Gorski (2009) emphasizes that teacher knowledge needs to extend beyond subject matter and include knowledge of societal issues such as multiculturalism. Holden and Hicks (2007) also challenged educators to incorporate global education that crosses national boundaries. In order to be global educators that are aware of societal issues, teachers must have opportunities to extend their learning beyond their classroom, school, districts, and cities. Each case within this study demonstrates teacher learning with regard to culture and globalization.

Mrs. Yang and Ms. Ko encountered cultural learning and global learning opportunities through different types of interactions and during different times of their journey. Both teachers had misconceptions regarding their partners' culture as it pertains to life and schooling. Throughout the process of engaging in a RLP, the teachers gained awareness of these misconceptions and reconstructed their knowledge using authentic, first-hand information. Within Mrs. Yang and Ms. Ko's RLP, Ms. Ko gained cultural knowledge for teaching through her students' interactions with Mrs. Yang's students.

While reading through the student letters, she came to recognize that she had held misconceptions about Chinese students. She had been wrong in believing that all Chinese students were good at math, liked math, were learning at the same level, and were learning in the same way. From the letters, she saw that the students in Mrs. Yang's class achieved at different levels, and while many liked math, there were many students who did not. Ms. Ko's initial perspectives are similar to the homogenized stereotypes outlined by Ryan and Slethaug (2010).

Furthermore, through conversation, both teachers gained awareness that student motivation to learn mathematics is an issue in the educational system of both nations. Ms. Ko, who had equated Chinese students' reputation for excellence in mathematics with high motivation for learning, was surprised when Mrs. Yang shared that she feels that her students are not very intrinsically motivated due to their hectic study schedules.

When first watching videos of Ms. Ko teaching, Mrs. Yang in China was initially shocked by the diversity in the classroom. She had travelled to Canada prior to this partnership and had misconceptions of the cultural diversity in the nation. Furthermore, through interactions with Ms. Ko, the Chinese teacher learned that Canadian classrooms are comprised of students from dozens of different countries and cultures with different religious beliefs. She was able to relate this to China's demographic of many minority cultures and was interested in how the teacher addressed these differences. Mrs. Yang experienced the need to accommodate different religious practices when she had to take into account gender while pairing students for letter writing.

In addition to learning about teaching and learning, the teachers also gained globalized knowledge of teacher education, teacher professional development, and school systems. The teachers within the RLPs were able to gain knowledge of learners through stories that their partners told when sharing teaching strategies, watching teaching videos, and observing and teaching in their partner's classrooms. Knowledge of the learners that are from a different cultural and schooling context is important because, as Vygotsky (1978) suggests, social and cultural goals need to be integrated into pedagogy.

By giving teachers an opportunity to see learners in a new and different context, it allows them to expand their understanding of elementary students. Furthermore, by comparing and contrasting the difference between learners in their own class to learners in their partner's class, the teachers are able to gain a better understanding about how context and culture can affect learners. This area of learner knowledge is directly related to knowledge for globalization and knowledge of multiculturalism for teaching outlined by Holden and Hicks (2007) and Gorski (2009).

When watching video of each their partner's teaching, the teachers were very attentive to the students' reactions to the teachers. Mrs. Yang found that watching the students in Ms. Ko's class was a big source of learning for her as she noticed that they were very focused on Ms. Ko when sitting on the carpet. She reflected upon her preconceptions of creating the ideal learning space for young learners and wondered if the structured desks and groupings in her own classroom were, in fact, beneficial or if the students would benefit from more freedom and movement during their learning.

Furthermore, Mrs. Yang points out that, when working in groups, the students have more freedom with regard to the roles they take and who they work with. Mrs. Yang suggests that, while this may not allow more the most efficiency or equal learning opportunity, she does see the merit in this type of group work as it is more similar to real life. Through comparison of her own classroom with Ms. Ko's, Mrs. Yang has done what Kelly (2013) describes as "making the strange familiar and the familiar strange" (p. 416), as she now questions her own familiar classroom's effectiveness.

Teacher confidence and their belief that they have the ability to positively impact student learning are essential to teacher growth and learning (Protheroe, 2008). Through engaging in the RLPs, both teachers became more aware of their own confidence and abilities. When entering into the RLP, Ms. Ko questioned her ability as a mathematics educator wondering if she had anything to contribute to their RLPs as her partners were math specialists. Through personal reflections and interactions with her partners, Ms. Ko gained confidence in her abilities and belief that she was able to contribute to the RLP and increase their partner's knowledge.

An example of increased teacher efficacy is seen with Ms. Ko as she reflects upon her RLP, sharing that she realized her own abilities and strengths as a math teacher when she was forced to share her knowledge with another person. She realized her lack of confidence was in her own head and that she had to push through her insecurities. Furthermore, Ms. Ko became aware of the importance of teacher efficacy when she watched a video of Mrs. Yang teaching. She had noticed that Mrs. Yang's confidence was captivating for her students and that they were so focused on her. Furthermore, she stated that it seemed like Mrs. Yang was having fun and that was important too. Ms. Ko later went on to combine this discovery with her prior goals of making sure math accessible to all her students.

Teacher efficacy increased when teachers learned that the RLP was utilizing the PCK that they had shared. When the teachers were taken out of their comfort zones and were challenged with teaching tasks that they had previously felt were beyond their ability, their success increased their self-efficacy.

8 Conclusion

Seeing the products of the RLP is an effective way for participants to gain motivation and confidence in their contributions and learning. The two main products that are produced by the RLP in this study that increased motivation are students' interactions and the lessons that integrate partner resources.

The RLP involved student interactions throughout the 26 months of this study. Mrs. Yang and Mrs. Ko continuously involved their students through a variety of different interactions that included learning about each other, sharing mathematics perspectives, and challenging each other with mathematics problems. In fact, Ms. Ko attributes student interactions to much of her learning throughout the RLP. This connection between the RLP and the teachers' work with students is a component that Darling-Hammond and McLaughlin (1995) suggest is essential to effective professional development.

In addition to student interactions, the RLPs gained momentum and the teacher gained confidence when resources or ideas that were shared were put into use by their partners. This was evident in Mrs. Ko's reflection on her own initial lack of confidence and later realization that she did have things to contribute after reading her partner's lesson study interviews. Making sure that teachers are aware of the impact their contributions make on the partners and on the RLP is important as very often resources are shared, used, and then participants are eager to move onto the next topic. The RLP facilitator plays a key role in providing teachers with prompts to discuss their use of shared resources and to prompt the sharing of teacher reflection.

We can conclude that, when teachers from Canada and China engage in a RLP, there are three main types of teacher knowledge learning that is shared: subject-specific teacher knowledge, pedagogical knowledge, and knowledge of the learner. When focusing on mathematics-specific teaching knowledge, the Canadian teacher was able to learn specific math content on topics such as algebra and multi-digit multiplication. The Chinese teacher focused their learning on the connections that the Canadian teacher drew between specific math content and other subjects such as geography, science, and language.

When sharing pedagogical knowledge, differentiated learning is of great interest to teachers in Canada and China. Recent curriculum reforms in both countries strongly urge teachers to differentiate for their individual learners resulting in teachers seeking out additional learning opportunities. Through cross-cultural reciprocal learning, the teachers in both countries are able to gain cultural knowledge of learners and gain insight into differentiating for learners with exceptionalities.

Although there are hundreds of thousands of Chinese immigrants in Ontario, it appears that teachers in both countries still have many misconceptions in terms of schooling, teaching, and learning in either country. These misconceptions can be challenged and corrected through teachers and students engaging in authentic cross-cultural interactions.

RLPs provide an environment for teachers to increase teacher efficacy. Within a safe environment, the teachers were able to develop trust and knowledge of their partner and facilitator, and teachers are able to engage in consistent and iterative self-reflection and comparison to identify areas of strength, growth, and the gaps in their knowledge.

References

Achinstein, B. (2002). Conflict amid community: The micropolitics of teacher collabora- tion. *Teachers College Record, 104*(3), 4212–4455.

Andrews, M. (2007). Exploring cross-cultural boundaries. In D. J. Clandinin (Ed.), *Handbook of narrative inquiry: Mapping a methodology* (pp. 489–511). Thousand Oaks, CA: Sage Publications Inc.

Arbough, F. (2003). Study groups as a form of professional development for secondary mathematics teachers. *Journal of Mathematics Teacher Education, 6,* 139–163.

Ball, D. L., Thames, M. H., & Phelps, G. (2008). Content knowledge for teaching: What makes it special? *Journal of Teacher Education, 59,* 389–407.

Ben-Peretz, M. (2011). Teacher knowledge: What is it? How do we uncover it? What are its implications for schooling? *Teaching Teacher Education, 27,* 3–9.

Connelly, F. M., & Clandinin, D. J. (1988). *Teachers as curriculum planners: Narratives of experience.* New York: Teachers College Press.

Connelly, F. M., & Clandinin, D. J. (1990). Stories of experience and narrative inquiry. *Educational Researcher, 19*(5), 2–14.

Connelly, F. M., & Clandinin, D. J. (2006). Narrative inquiry. In J. Green, G. Camilli, & P. Elmore (Eds.), *Handbook of complementary methods in education research* (pp. 375–385). Mahwah, NJ: Lawrence Erlbaum.

Connelly, F. M., Clandinin, J. D., & He, M. F. (1997). Teachers personal practical knowledge on the professional knowledge landscape. *Teaching and Teacher Education, 13*(7), 665–674.

Connelly, F. M., & Xu, S. (2015). Reciprocal learning: Comparative models and the partnership project. In *Reciprocal learning & symbiotic development* (pp. 1–18). Shanghai.

Creswell, J. (2012). *Educational research: Planning, conducting, and evaluating quantitative and qualitative research* (4th ed.). Upper Saddle River, NJ: Pearson Education.

Darling-Hammond, L., Hyler, M. E., Gardner, & Espinoza, D. (2017). *Effective teacher professional development.* Palo Alto, CA: Learning Policy Institute.

Darling-Hammond, L., & McLaughlin, M. (1995). Policies that support professional development in an era of reform. *Phi Delta Kappan, 76*(8), 597–604.

Desimone, L. M. (2011). A primer on effective professional development. *The Phi Delta Kappan, 92*(6), 68–71.

Fox, C. F. (2006). Stories within Stories: Dissolving the boundaries in narrative research and analysis. In S. Trahar (Ed.), *Narrative research on learning: Comparative and international perspectives* (pp. 47–60). Oxford: Symposium Books.

Gojmerac, A., & Cherubini, L. (2012). Connecting for a greater purpose: An examination of a district professional learning community. *Journal of Teaching and Learning, 8*(1), 43–58.

Gorski, P. (2009). What we're teaching teachers: An analysis of multicultural teacher education coursework syllabi. *Teaching and Teacher Education, 25*(2), 309–318.

Guskey, T., & Huberman, M. (1995). *Professional development in education: New paradigms and practices.* New York: Teachers College Press.

Holden, C., & Hicks, D. (2007). Making global connections: The knowledge, understanding and motivation of trainee teachers. *Teaching and Teacher Education, 23*(13), 13–23.

Jao, L., & McDougall, D. (2016). Moving beyond the barriers: Supporting meaningful teacher collaboration to improve secondary school mathematics. *Teacher Development, 20*(4), 557–573.

Kelly, P. (2013). Comparative pedagogy: Making sense of cultural complexity. *Research in Comparative and International Education, 8*(4), 415–427.

Nelson, T., & Slavit, D. (2008). Supported teacher collaborative inquiry. *Teacher Education Quarterly, 35*(1), 99–116.

Protheroe, N. (2008). Teacher efficacy: What is it and Does it matter? *Principal, 87*(5), 42–45.

Ryan, J., & Slethaug, G. (2010). *International education and the Chinese learner*. Hong Kong: Hong Kong University Press.

Schlager, M., & Fusco, J. (2004). Teacher professional development, technology, and communities of practice: Are we putting the cart before the horse? In S. Barab, R. Kling, & J. Gray (Eds.), *Designing virtual communities in the service of learning* (pp. 120–153). New York: Basic Books.

Shulman, L. S. (1987). Knowledge and teaching foundations of the new reform. *Harvard Educational Review, 57*(1), 1–22.

Stake, R. E. (1995). *The art of case study research*. Thousand Oaks, CA: Sage.

Tye, K. (1999). *Global education: A worldwide movement*. Orange, CA: Independence Press.

Vygotsky. L. (1978). Interactions between learning and development. In *Mind and Society*. Cambridge, MA: Harvard University Press.

Wilson, S. M., & Berne, J. (1999). Teacher learning and the acquisition of professional knowledge: An examination of research on contemporary professional development. *Review of Research in Education, 24*, 173–209.

Xu, S. J., & Connelly, F. M. (2013). *Reciprocal learning in teacher education and school education between Canada and China* (*2013–2020*). Partnership Grant Proposal Package. Social Sciences and Humanities Research Council of Canada (SSHRC) Partnership Program.

Yazan, B. (2015). Three approaches to case study methods in education: Yin, Merriam and Stake. *The Qualitative Report, 20*(2), 134–152.

CHAPTER 7

Mathematics Teachers' Perspectives on Effective Learning Through Reflections on Their Experiences in a Canada–China Learning Partnership

Ying Chen

1 Introduction

Increasing demand for student learning and achievements has pressured teachers to improve their teaching practice through professional learning, because the quality of teacher learning directly shapes the learning and achievements of students (Vescio, Ross, & Adams, 2008). However, with the blossoming of teacher professional learning, problems emerged in terms of its content, format, and procedure. Despite continuing efforts to reconsider and innovate in teacher learning (Borko, 2004), the situation has hardly improved. Consequently, instead of offering a positive impact on school practices, professional learning became a burden for

Y. Chen (✉)
Ontario Institute for Studies in Education, University of Toronto, Toronto, ON, Canada
e-mail: amandaying.chen@mail.utoronto.ca

S. C. Zhu et al. (eds.), *Reciprocal Learning for Cross-Cultural Mathematics Education*, Intercultural Reciprocal Learning in Chinese and Western Education,
https://doi.org/10.1007/978-3-030-56838-2_7

many teachers. To address this issue, many researchers (e.g., Goodson, 1991; Kirk & MacDonald, 2001; Kooy & van Veen, 2012; van Veen & Kooy, 2012) emphasized the importance of including the voices of teachers in all aspects of their learning and suggested that this was the most effective way to construct knowledge toward improved learning for both teachers and students.

Many recent studies (e.g., Green & Whitsed, 2013; Little, 2012; van Veen, Zwart, & Meirink, 2012; Webster-Wright, 2009) have criticized traditional programs, and their failure to meet teachers' learning needs in evoking knowledge growth or improving their practices. In such misconstrued learning programs, teachers felt ignored as their professionalism is not recognized (Kooy & Colarusso, 2012) or was diminished. They felt mistreated as passive recipients of knowledge in a learning process that focused on abstract skills and content (Loughran, 2012) rather than recognizing their personal knowledge and experiences. Typically, teachers felt offended when so-called experts, who lacked the teachers' daily knowledge of both teaching and students, lecture them. Under these circumstances, the teachers' autonomy was neglected, and they were not given the power to make decisions or choices in their own learning or teaching.

Pedder, James, and Macbeath (2005) addressed several problems in teachers' values of professional learning, which they blame on the lack of institutional support in freeing professional learning. The role of research, collaboration, reflection, practice, student voice, and dialogue on professional learning for teachers, which were found to be the most critical among factors in improving teachers' learning based on the review of over 30 years of research into teachers' learning, were not recognized nor valued by many teachers. If the use of research was widely supported by schools, "if the conditions of their work [were] conducive to the development of trust and openness" (p. 235), "if the environments that they [were] working in are congenial to collective learning" (p. 236), if they [were] free to change what they do with their knowledge recognized, teachers tend[ed] to have more "resilience and self-confidence to take the risks involved" (p. 236) in professional learning.

The study of experience is a path to constructing knowledge (Ackermann, 2004). Situated in a Canada–China Reciprocal Learning Project, this study, through an in-depth investigation of the experiences of teachers, addresses the influential factors, existing problems, and challenges for teacher learning. To that end, it aimed to advocate a more teacher-centered environment for professional learning in schools, as well as to encourage teacher voice and empowerment in the field of curriculum studies and teacher development.

2 Conceptual Foundations

To improve the contexts of teacher learning, researchers have previously defined a few key principles of transformative professional learning. For example, effective teacher learning engages teachers with the goal of reconstructing foundations in school-based environments (Lieberman, 1995; McLaughlin & Talbert, 2006). Meanwhile, it could provide a learning space for the interactive process of dialogical learning through critical reflection and collaborative problem-solving (Putnam & Borko, 2000) over a sustained period of time (Darling-Hammond & Richardson, 2009). Nevertheless, shared beliefs about how to build transformative contexts of learning did not necessarily lead to consistent results for teachers. In many cases, professional learning sessions failed to either take full advantage of teachers' practical experiences or bring rewarding learning outcomes, because the hierarchical structure of learning restricted teachers' full participation in leading positions (Wallace, 2003).

Commonly, effective learning contexts were those in which teachers were considered professionals and intellectuals (Giroux, 1988) capable of determining their learning needs and with the confidence to identify their active and leading roles in teaching and learning (Graven, 2004). Other studies had also found that, in effective social relationships among peers, teachers showed commitment and emotional engagement in openly sharing learning skills and resources with trust and equity (Day, Elliot, & Kington, 2005). With shared goals in those learning contexts, teachers took risks to improve practices and resolve struggles (Lieberman, 1995) and actively created learning opportunities to make transformative changes within and around themselves (Barth, 2007).

2.1 *Social Theories of Learning*

Social and meaningful education (Vygotsky, 1978) engages learners voluntarily in collaborative activities that allow them to "interact with others in specific problem-solving environments" (Moll, 1990, p. 11).

Learning as a social practice emphasizes learning that took place in interaction with others, not in isolation. It required learners to act as the initiators of learning, collaboration and interaction as the key media of learning, and shared learning goals. Dewey (2010), along with Vygotsky (1978), whose democratic learning theory had a remarkable influence on education, also proposed to free and center learners in the process of social learning, as he defined democratic learning as new learning that encourages creative thinking and interactive collaborations.

To achieve this democratic social learning, Bakhtin's (1986) dialogical learning theory emphasized using dialogues to bridge multiple voices for communicating thoughts, stimulating thinking, and constructing knowledge. In social learning contexts, engaging in meaningful interactive talk with each other allowed learners to make sense and to verbalize what was in their minds (Butt, 1989). As Mercer (2000) also stated, the course of argument among different minds was necessary for constructing new knowledge from making sense of previous knowledge. Through the dialogue with others, the processing of inner voices, and the interaction of multiple voices, the process of arguing affords new knowledge construction.

2.2 *Teacher Learning in Collaboration*

Over the years, the value of collaborative learning has been increasingly explored and advocated for learning in schools (Vangrieken, Dochy, Raes, & Kyndt, 2015). Studies show that collaboration in classroom teaching became a key factor for student success (Egodawatte, McDougall, & Stoilescu, 2011), improved student understanding (Wigglesworth, 2011), and student learning experiences (Main & Bryer, 2005). As a result, collaboration started to be widely applied in all kinds of learning contexts, regardless of the subjects. Moreover, learning in collaboration was increasingly implemented and researched in the field of teacher learning. With the high turnover in some school districts, helping teachers to find professional support through collaboration may keep some teachers in school longer. For example, Berry et al. (2009) found that collaboration led to more effective teaching and that teachers were more likely to stay in secondary schools. Teacher collaboration was also helpful to improve teacher practice (Westheimer, 2008).

Because of the learned value of collaboration, many organizers for teacher programs advocated for teacher collaboration and put forward different modes of learning that involve collaboration to some extent (Egodawatte et al., 2011). Some of these learning programs engaged teachers in collaborations that lead to an improvement in teacher efficacy, productive teacher conversations, and a cultural shift to more equity (Slavit, Kennedy, Lean, Nelson, & Deuel, 2011).

Some research criticizes collaborative or community-based teacher learning as they caused negative impact on teachers (Darling-Hammond & Richardson, 2009; Dillengbourg, 2002; Dufour, 2004). Due to an inadequate understanding of the nature of collaborations, the increasing favor of collaborative learning became a concern, as researchers found it's being overused, and applied lightly in standardized ways that would not bring positive impact on learning. For example, Johnson (2003) noticed an increase in teacher competitiveness, interpersonal conflicts, and a loss of autonomy. As a result, without gaining a greater understanding of the true nature of collaboration, some teacher-learners developed a negative impression of collaborative learning. They felt that their dissonance voices were silenced, and they had to conform to the norms of the majority (Johnson, 2003). There was also recognition that different teaching styles and pedagogical beliefs were intensified at the group level (Main & Bryer, 2005).

Because many ineffective teacher learning approaches have been labeled as collaborative learning, teachers started to lose faith in collaboration (Gajda & Koliba, 2008). Helping teachers understand the nature of collaboration and its value in teaching and learning became an urgent issue in the field of professional learning.

3 Research Methodology

This study is contextualized within an ongoing longitudinal Canada–China Reciprocal Learning Partnership (Xu & Connelly, 2013). The partnership project consisted of five research teams: Mathematics Education Team, General Education Team, Science Education Team, Language Education Team, and a Teacher Education Team. This chapter focuses on teacher learning of one of the teacher partnerships in the Mathematics Education Team, and so the teacher partners focused on the teaching and learning of mathematics in their partnership activities.

Because of the value of individualism, I chose case study methodology, which used the voice of teachers extensively to construct meanings of teaching and learning. Instead of generalizing the findings, I used specific examples of each teacher to describe teacher's individual and personal ways of learning and thinking.

Narrative served as the most functional methodological framework to construct, analyze, understand, and present teacher experience in its wholeness (Polkinghorne, 1995). Using teachers' narratives afforded better understanding of teachers' ways of knowing. To construct authentic knowledge for teacher learning, I gathered teachers' narratives on teacher learning to construct meanings of authentic learning for teachers. It contributed to improving professional learning contexts that provided space for teacher voices and acting in innovative roles. I used direct quotes from each teacher to present authentic thoughts from them, trying to eliminate the changing of words according to my personal interpretations. Through the conceptual foundation as well as the research design of this study, I suggest broadening the inclusion of teacher voice in curriculum studies and to apply research methodologies that allow teacher voice to be presented to wider audiences.

In this case, we first worked as a research team to identify the goals of the project and the possible schools to approach to discuss the focus of the project. We selected teachers from an Ontario school, who was already part of another research project. In China, the research team approached a participating research school to ask for teacher volunteers. We paired up teachers (one from Toronto, Canada, the other from Changchun, China) who taught the same grade.

This chapter presents a case study that focuses on a Canada–China learning partnership that centered on a pair of teachers. Sabrina is a Canadian teacher from Jamaica and was teaching grade six mathematics, science and language arts. Yang was a grade six teacher of mathematics. In terms of the participatory activities of learning, Yang tended to consult her principals or the administrators first. The advantages and limitations of this tradition arose and impacted her teacher learning, and the ways that she interpreted learning. This study used teacher voices to present their different thoughts on teacher learning in the partnership.

3.1 Data Collection Methods

Data were gathered from a range of sources: teacher interviews, field notes after each school visit, classroom observation, online conversations, videotaped lessons, videotaped teacher Skype meetings, and exchanged documents including student work, lesson plans, and school booklets of history and curriculum. The teacher interview was the main way of collecting data for this research. A series of interviews were conducted throughout the study, beginning in Fall 2015, to understand the background of the teachers in education and teaching, their motivations to participate in the project and their expectations, as well as their visions of teaching and learning. Then, after a year of participation, the teachers were interviewed to reflect on their experiences in Canada–China Reciprocal Learning Project, and their suggestions for improving the project to better meet their learning needs.

In March 2017, three months before they ended the two-year cycle of reciprocal learning partnership, the teachers were interviewed regarding their previous experiences of professional learning. The interview followed up on topics emerging from conversations that had occurred during the reciprocal learning partnership. They were also asked questions regarding their perceptions of the challenges that they faced during the learning partnership and how they had overcome them. The interviews contained semi-structured questions and were conducted either face-to-face or online via Skype. The length of each interview was between fifteen minutes and one hour.

4 Effective Learning Experiences for Teachers

A community of learning was built over the two-year partnership, as the study engaged teachers, researchers, principals, and students in dialogical communications on teaching and learning, through online meetings and lesson exchanges.

4.1 *Effective Learning Achieved Through Dialogues on Teaching in Classrooms*

Teachers met over Skype monthly as the main activity for teacher exchange in the partnership. The agendas of each meeting were prepared ahead of time in English and Chinese for each teacher. At each meeting, at least one researcher was present to help teachers communicate with their partner in the other language. In those meetings, teachers expressed their inquiries of interest, exchanged different opinions, and discussed them in depth. Teachers talked with each other in detail regarding different educational policies and their school cultures. They also shared their knowledge and experiences on teaching specific mathematics content and exchanged strategies that they applied in teaching in their daily work.

Meanwhile, teachers watched their partner's lessons and, afterward, raised questions to enhance the understandings of their partner's practices, as well as to provide feedback by reflecting on their own. For such exchange, Sabrina taught a lesson on probability and Yang taught a lesson on ratio. In order to provide an understandable watching experience, we added captions to those videotaped lessons using iMovie, either in English or in Chinese.

The teachers commented positively on the critical role that dialogues played in their partnership exchanges. Yang commented that meeting online was "the most touching part of reciprocal learning, even though we were so far away, cross-national, we got to communicate online one-to-one…which I never had before." Those meetings brought both teachers together and closer. Immediately, they found similarities and relevance in their teaching regardless of different cultures, and they started sharing with each other openly. Meanwhile, Sabrina thought that this type of communications through lesson watching was most valuable to her in this partnership. She said, "I value the best practices, the pedagogy, and the approach to instruction that I had seen." She elaborated on three aspects on which the videotaped lessons had benefited her teaching: student behavior, student engagement, and differentiated learning. Sabrina commented that those were the main challenges in her classroom, and she found helpful strategies in the Chinese lessons.

Through the lesson exchange, both teachers observed and learned about the classroom teaching of the other partner. The videos presented teachers with the actual setting of each classroom. Based on close observations, teachers engaged in the active reflection of their own perspectives on teaching a specific lesson, compared the similarities and differences of their teaching with the other partner, as well as provided suggestions for alternative strategies to their partners in improving the lesson. By exchanging the feedback between partners, the teachers learned about the different classroom cultures, styles of teaching, visions and beliefs of teaching and learning in Mathematics.

4.2 *Effective Learning Achieved Through Principal Involvement*

Yang was scheduled to travel to Canada to meet in Sabrina's school as part of the exchange. Unfortunately, due to visa refusal, Yang was unable to visit Toronto in person, so she missed the school visit to Sabrina's class. However, her principal came to Toronto as a representative and visited Sabrina's school. This school visit was a successful part of the partnership between Sabrina and Yang, because it deepened their partnered friendship and encouraged both teachers' learning.

The presence of Yang's principal showed Sabrina the respect and attention from Yang's school to this partnership. She was encouraged by the fact that educators came from China to visit her in person and to learn from her. On the day of visits, Sabrina welcomed these visitors and accompanied them energetically throughout the school tour. Through this experience, Principal Gang was able to experience learning in Ontario in person and took away lots of valuable information regarding Canadian schools for Yang and also for other teachers and students in her school.

Both teachers consistently and respectively commented that they believed that successfully engaging in different interested parties in the process of teacher learning made learning more effective. From the beginning of the partnership, the teachers expressed their expectations to involve other stakeholders including administrators, colleagues, students, and parents, as part of their learning exchange. However, because they had different school contexts, the type of support they received was distinguished, which led to various experiences of learning.

Active participation of Principal Gang in the partnership made a difference for Yang's learning. Among the two teachers, whenever they were talking about her own teaching and learning, Yang always commented on the substantial support she received from her principal. I learned from Yang that Principal Gang was a very selfless educator because, as a principal, she always managed to meet the needs of her teachers and students. She did not keep a distance from others in schools. Instead, she constantly met with individual teachers and students to obtain feedback and suggestions on improving schools and student achievement. She respected opinions from teachers and students as well as encouraged them to think creatively and act boldly. Yang said that all of those qualities were exceptional for a principal in China.

As one of the model principals in her province, Principal Gang was often invited to talks and interviews from the local TV stations. Her professional journey and visions on education were published in newspapers and journals. As a principal, Principal Gang always looked out for opportunities for her school to grow, such as this partnership. Principal Gang pursued this opportunity for her school and found a responsible teacher to carry it on, hoping to achieve benefits beyond those for one teacher. Yang clearly indicated her appreciation of having the trust from Principal Gang to participate in this partnership.

4.3 Effective Learning Achieved Through Student Engagement

Students were the other stakeholders that were actively involved in this partnership. The association between Sabrina and Yang involved their students by having them videotape and exchange personal greetings and questions. Yang chose six students from her class to videotape their individual greetings and sent the videos to Sabrina's class. In return, Sabrina received six student volunteers to respond to Yang's students in the mode of videotaped conversations. From the beginning of the research, both Sabrina and Yang expressed interest in involving their students in communication as part of their learning partnership. Their intention was to start communication between their students so that to improve the overall partnership experience.

Through this exchange, students had an opportunity to see the students from a different country to express their thoughts on mathematics learning, to comment on their mathematics teachers, and to inquire about another culture. This benefited the teachers, too, as they heard about their students' perspectives, attitudes, and needs in classroom learning. For example, Sabrina reacted very emotionally while watching her students talking about mathematics learning and commenting highly on herself regarding assisting students' learning. After watching the videos, Sabrina was very proud of the openness of her students, and she said:

> First, I appreciate their sincerity in what they expressed. I am glad that they had the opportunity to participate in this project because the students in China can know better and understand what their experiences are in Canada…It is good for them to learn about each other… any biases that they may have [might start to change because of this partnership] …just to let them have a stronger connection with another class, which is associated with my learning partnership.

4.4 *Teachers' Opinions on Effective Learning for Teachers*

4.4.1 Sabrina

Toward the end of the two-year partnership, I interviewed Sabrina for reflections on her experiences in this partnership. I asked her to compare her previous experiences of learning to the current partnership as well as to evaluate the main components of the partnership to make learning more efficient for teachers. Compared to her previous experiences of learning, Sabrina considered them very similar in terms of the ways that she learned, and the learning outcomes in informing her classroom practices to improve student learning. Of all the learning activities in this partnership, Sabrina evaluated the lesson watching as one of the most efficient types of learning for her. Through the lessons from her Chinese partners, she learned different strategies to enhance student learning, and she was motivated to challenge herself in testing different approaches continuously.

Sabrina also commented that the context of teaching in China was very different from hers. She observed that the Chinese students were comparatively more receptive of knowledge and had more manageable behaviors in class. On the contrary, it was one of the biggest challenges for her to manage a classroom of students performing at different levels of ability, even though she continued to make efforts in improving on this.

On reflecting upon her learning experiences in the partnership, Sabrina gave positive comments: "I have since embraced the reciprocal learning project. I have found that I am even more open, more receptive of new challenges." Sabrina thought the learning did benefit herself with long-term impact and also others: "My students benefit anyways, but my colleagues, too because I share with them what I have learned, what I have experienced, while navigating this journey." Sabrina expressed that she "wanted to continue with this project as long as possible" because she wanted to continue learning from her partners. Sabrina believed that the future held promise. She looked forward to learning more from her Chinese partners to challenge herself, in order to improve her students' learning:

> It really helps me to expand more on what I have to share with my students. I have been challenged to be more creative, be more dynamic, be even more flexible. Just allowing the students to move more at their

> own pace, as opposed to the pace I would like for them to move. After all, learning is not something that can be forced. They just have to learn the best way that they can, at the pace that they are able to. It is overall a very positive experience for as long as I can remain in this progress as I mentioned I hoped to.

Sabrina had been thinking of sharing her lessons from the partnership with her colleagues in the school. Through the partnership, she became more sensitive to cultural factors in the context of learning and teaching. She hoped to make a difference in her school by "encouraging teachers to include more of the cultural aspect of learning in their mathematics lessons." She believed that "even if the students are from different countries, or culturally, things are different at home. [It is worth considering] …how can culture be infused in Mathematics class."? In this way, "students may see more of themselves represented, they have more of a voice, give them that voice in the Mathematics lesson."

Sabrina identified some of the challenges for her students. She used one example of her ELL students, "who probably feel somewhat ostracised because they do not see their culture, they do not see themselves in the mathematics lessons, in the textbooks." She imagined that, "by allowing them that opportunity, that platform, to showcase what they understand, what they have learned to the rest of the class. That is something else."

She also saw the benefits for teachers through this teaching experience: "Just empowering other teachers, who may find Mathematics to be a challenge. Because there are some teachers who think they would rather teach areas in which they have their teaching qualification." She explained, "Of course, I understand, but it is also important for them not to feel that they will not be able to function, or to be effective in another subject area. Because sometimes they end up surprising themselves, that they excel as opposed to what they anticipated."

4.4.2 Yang

The reason for Yang's participation in the project was that, as the opportunity was announced, and her principal contacted her, she expressed a strong will and passion for participating. She thought, through this learning, she could bring an eye-opening experience for herself as well as her students.

Second, she personally wanted to accept this task and succeed. Yang considered herself very open to learning new things. She claimed that anything that could be beneficial to herself, or her students, or even her whole life, she would like to know and adjust to the possibilities. She had the will to pursue this goal herself. Compared to her previous learning experiences, Yang commented on her experience in this partnership below:

> Similarly, I learned new perceptions of education. I always have takeaways. Canadian teachers have opportunities to come to my school to learn about my teaching surroundings, and my students. This part I feel very intimate. Comparatively, I like such form of learning better because teachers have interactions. Most teacher training is a one-way format like monologues. In such way, we always listen to, take notes, and learn things. They rarely, unlike reciprocal learning activities, provide opportunities for us to face each other, to talk about, to act out, or to present our work. I like such interactive ways. Personally, I have gained a lot.

The overall experience was very positive for Yang. She thought it was a unique experience of learning that she really appreciated. Originally, she did not expect to be able to talk to, or to share about her life, or teaching with her partner one-to-one, face-to-face, in detail. She felt connected to her right away after the first online conversations. What she valued most of the partnership was that, "it infused the teaching inside of China and outside of China. It was eye-opening for both teachers and students." In the process, she was devoted to communicating with her partner, and she believed that, her "level of teaching has been elevated through such process of learning."

Yang was grateful for being part of this valuable study and of the learning that was accomplished through different types of learning exchanges through activities and the developing relationships between teachers that evolved over the course of the project. She thought that the research leaders of both sides, who created this opportunity of learning, and facilitated the project, were one of the factors that made it successful; and the second important factor was the individual teachers involved that continuously made efforts to bring new things to the classrooms from this partnership.

Reflection on her journey of professional learning and growth since her participation in reciprocal learning from 2015, Yang had very positive comments:

> From not quite understanding the project, the activities, always asking you guys about the next steps, to gradually think on my own. Every time after communicating with you and Ms. Sabrina, I feel it opened my mind and started to bring what I learned back to my classroom. Now I am confident enough to use what I saw and learned from Canadian teachers to improve my own teaching. Over the 1.5 years, I have learned to think independently and accomplish the activities with more confidence.

Through the partnership learning experiences in the project, Yang became more independent in decision-making for her own learning, as well as in choosing teaching resources for her students. She also admitted that the partnership learning experiences had influenced her profession, as well as roles in her school.

> Previously, my class was more traditional. I rarely involved my students in presenting their own thoughts. It was more teacher-centered. Students rarely had opportunities to finish expressing their thoughts and completing a task on their own before teachers drew a conclusion. I used to impose my thoughts on students before they explained the process of their thoughts. However, now I tend to learn together with my students, infusing their thoughts with mine.

Based on her comments, Yang had improved her teaching practices through the partnership. Since then, she developed more respect for her students' thoughts and has been trying to shift her lessons from teacher-centered to student-centered.

Envisioning the sustainable benefits of reciprocal partnership learning, Yang said that her wish was to continuously "be inspired by the reciprocal learning experience." Specifically, she would "continue applying the concept of the student-centered classroom in the teaching." She kept in mind that "teachers are facilitators, and the students are the centre of the class." Therefore, "the takeaways from reciprocal learning will be the leading concept for her, and even for her country," she said.

5 Discussion

The study shows that effective teacher learning was achieved through interactive dialogues between teachers around their experiences and perceptions of teaching and learning. The partnership respected teachers' perspectives and allowed teachers to converse with each other through direct conversations, with extended engagement of principals, parents, students, and researchers. Through reflecting via in-depth dialogues, the teachers negotiated existing knowledge and perceptions and constructed new ones for change in practices. The differing amount of involvement of other stakeholders influenced the effectiveness of teacher learning for each respective teacher. In the final reflections, the teachers evaluated their learning effectiveness, which implied suggestions in improving learning for teachers.

The study exemplified the value of respecting teacher voice in educational research. The birth of the study was rooted in the advocacy of teacher voice. The study set the stage for teachers to contribute their personal voice.

International learning opportunities are valuable for teachers but certain unavoidable challenges prevent teachers' involvement in cross-cultural contexts of learning. Teacher learning in cross-cultural contexts in this study presented various challenges for both teachers. The case-specific challenges included language barriers, time difference between two countries, the incompatible tools of communications, and cross-border travel restrictions. Being aware of the differences in the contexts of Canada and China, in terms of the background of teacher learning, I believe that studying in such a reciprocal learning project contributed to the knowledge base of supporting effective learning for teachers to collaborate in international settings with deeper understanding of cultural influences on teacher learning.

Among many challenges, language barrier influenced the ways of the learning for teachers. In this study, direct dialogical communications among teachers were limited because they were dependent on the facilitators to achieve basic communication with the other partner. Teacher participants expressed that they hoped that they could speak the other language so that they could understand their partner better and express themselves directly without being translated by another person.

References

Ackermann, E. K. (2004). Constructing knowledge and transforming the world. In M. Tokoro & L. Steels (Eds.), *A learning zone of one's own: Sharing representations and flow in collaborative learning environments*. Amsterdam: IOS Press.

Bakhtin, M. M. (1986). *Speech genres and other late essays* (V. McGee, Trans.). Austin: University of Texas Press.

Barth, R. S. (2007). The teacher leader. In R. H. Ackerman & S. V. Mackenzie (Eds.), *Uncovering teacher leadership* (pp. 9–36). Thousand Oaks, CA: Sage.

Berry, B., Daughtry, A., & Wieder, A. (2009). *Collaboration: Closing the effective teaching gap*. Retrieved from Centre for teaching quality website: http://www.teachingquality.org/content/collaboration-closing-effective-teaching-gap.

Borko, H. (2004). Professional development and teacher learning: Mapping the terrain. *Educational Researcher, 33*(8), 3–15.

Butt, D. (1989). *Talking and thinking: The patterns of behavior* (2nd ed.). Oxford: Oxford University Press.

Darling-Hammond, L., & Richardson, N. (2009). Teacher learning: What matters? *How Teachers Learn, 66*(5), 46–53.

Day, C., Elliot, B., & Kington, A. (2005). Reform, standards and teacher identity: Challenges of sustaining commitment. *Teaching and Teacher Education, 21*(5), 563–577.

Dewey, J. (2010). *Democracy and education: An introduction to the philosophy of education*. London: Indo-European.

DuFour, R. (2004). What is a "professional learning community?". *Educational Leadership, 61*(8), 6–11.

Egodawatte, G., McDougall, D., & Stoilescu, D. (2011). The effects of teacher collaboration in grade 9 applied mathematics. *Educational Research for Policy and Practice, 10,* 189–209.

Gajda, R., & Koliba, C. (2008). Evaluating and improving the quality of teacher collaboration. *NASSP Bulletin, 92*(2), 133–153.

Giroux, H. (1988). *Teachers as intellectuals: Toward a critical pedagogy of learning*. Granby: Bergin & Garvey.

Goodson, I. F. (1991). Sponsoring the teacher's voice: Teachers' lives and teacher development. *Cambridge Journal of Education, 21*(1), 35–45.

Graven, M. (2004). Investigating mathematics teacher learning within an in-service community of practice: The centrality of confidence. *Educational Studies in Mathematics, 57*(2), 177–211.

Green, W., & Whitsed, C. (2013). Reflections on an alternative approach to continuing professional learning for internationalization of the curriculum across disciplines. *Journal of Studies in International Education, 17*(2), 148–164.

Johnson, B. (2003). Teacher collaboration: Good for some, not so good for others. *Educational Studies, 29,* 337–350.

Kirk, D., & MacDonald, D. (2001). Teacher voice and ownership of curriculum change. *Journal of Curriculum Studies, 33*(5), 551–567.

Kooy, M., & Colarusso, D. (2012). The transformative potential of teacher and student voices: Reframing relationships for learning. In M. Kooy & K. van Veen (Eds.), *Teacher learning that matters: International perspectives* (pp. 80–99). New York: Routledge.

Kooy, M., & van Veen, K. (2012). *Teacher learning that matters: International perspectives*. New York, NY: Routledge.

Lieberman, A. (1995). Practices that support teacher development: Transforming conceptions of professional learning. *Phi Delta Kappan, 76*(8), 591–596.

Little, J. W. (2012). Professional community and professional development in the learning-centered school. In M. Kooy & K. van Veen (Eds.), *Teacher learning that matters: International perspectives* (pp. 22–46). New York, NY: Routledge.

Loughran, J. (2012). Professional learning: Creating conditions for developing knowledge of teaching. In M. Kooy & K. van Veen (Eds.), *Teacher learning that matters: International perspectives* (pp. 47–63). New York, NY: Routledge.

Main, K., & Bryer, F. (2005). What does a 'good' teaching team look like in a middle school classroom? In B. Bartlett, F. Bryer & D. Roebouck (Eds.), *Stimulating the "action" as participants in participatory research.* Proceedings of the 3rd International Conference on Cognition, Language, and Special Education. http://www98.griffith.edu.au/dspace/handle/10072/2538.

McLaughlin, M., & Talbert, J. E. (2006). *Building school-based teacher learning communities: Professional strategies to improve student achievement.* New York: Teachers College Press.

Mercer, N. (2000). *Words and minds: How we use language to think together*. London: Routledge.

Moll, L. C. (Ed.) (1990). *Vygotsky and education: Instructional implications and applications of sociohistorical psychology*. New York: Cambridge.

Pedder, D., James, M., & MacBeath, J. (2005). How teachers value and practise professional learning. *Research Papers in Education, 20*(3), 209–243.

Polkinghorne, D. E. (1995). Narrative configuration in qualitative analysis. *Qualitative Studies in Education, 8,* 5–23.

Putnam, R. T., & Borko, H. (2000). What do new views of knowledge and thinking have to say about research on teacher learning? *Educational Researcher*, 4–15.

Slavit, D., Kennedy, A., Lean, Z., Nelson, T., & Deuel, A. (2011). Support for professional collaboration in middle school mathematics: A complex web. *Teacher Education Quarterly, 38*, 113–131.

Vangrieken, K., Dochy, F., Raes, E., & Kyndt, E. (2015). Teacher collaboration: A systematic review. *Educational Research Review, 15,* 17–40.

van Veen, K., & Kooy, M. (2012). Stepping back and stepping in: Concluding thoughts on the landscapes of teacher learning that matters. In M. Kooy & K. van Veen (Eds.), *Teacher learning that matters: International perspectives* (pp. 255–260). New York, NY: Routledge.

van Veen, K., Zwart, R., & Meirink, J. (2012). What makes teacher professional development effective? In M. Kooy & K. van Veen (Eds.), *Teacher learning that matters: International perspectives* (pp. 3–21). New York: Routledge.

Vescio, V., Ross, D., & Adams, A. (2008). A review of research on the impact of professional learning communities on teaching practice and student learning. *Teaching and Teacher Education, 24*(1), 80-91.

Vygotsky, L. S. (1978). *Mind and society: The development of higher psychological processes*. Cambridge, MA: Harvard University Press.

Wallace, J. (2003). Learning about teacher learning: Reflections of a science educator. In J. Wallace & J. Loughran (Eds.), *Leadership and professional development in science education: New possibilities for enhancing teacher learning* (pp. 1–16). London and New York: RoutledgeFalmer.

Webster-Wright, A. (2009, June). Reframing professional development through understanding authentic professional learning. *Review of Educational Research, 79*(2), 702–739.

Westheimer, J. (2008). Learning among colleagues: Teacher community and the shared enterprise of education. In M. Cochran-Smith, S. Feiman-Nemser, & J. McIntyre (Eds.), *Handbook of research on teacher education* (pp. 756–782). Reston, VA and Lanham, MD: Association of Teacher Educators and Rowman.

Wigglesworth, M. (2011). *The effects of teacher collaboration on students' understanding of high school earth science concepts* (Master's thesis). http://etd.lib.montana.edu/.

Xu, S. J., & Connelly, F. M. (2013). *Reciprocal learning in teacher education and school education between Canada and China.* Unpublished research project proposal. An eight-year standard research grant funded by Social Sciences and Humanities Research Council (SSHRC) of Canada [Grant No. 895-2012-1011].

CHAPTER 8

Reciprocal Learning Between Chinese and Canadian Middle School Mathematics Teachers

Ying Chen

1 Introduction

International collaborations and partnership become increasingly popular with the continual development of technology offering flexibility and new opportunities for teacher-learners from distant geographical points to connect and communicate (Cogburn & Nanette, 2003). The role of international collaborations and partnerships potentially offers sustainable benefits for the global society (Goodwin, 2020) and, more importantly, sets the stage for cross-country dialogue likely to bring together various perspectives on knowledge construction and its mobilization worldwide (Cai & Knuth, 2011). International teacher partnership may engender in-depth knowledge exchanges between different countries while enhancing

Y. Chen (✉)
Ontario Institute for Studies in Education, University of Toronto, Toronto, ON, Canada
e-mail: amandaying.chen@mail.utoronto.ca

S. C. Zhu et al. (eds.), *Reciprocal Learning for Cross-Cultural Mathematics Education*, Intercultural Reciprocal Learning in Chinese and Western Education,
https://doi.org/10.1007/978-3-030-56838-2_8

cultural understandings on teaching and learning (Hayhoe & Pan, 2001; Moore, 2012).

The partnership in this study between Canada and China exemplifies a connection between the common binary of the West and East. These two sides have always been a myth to each other (Dewey, 1926; Said, 1978). Often the understanding between them is limited to media and stereotypes until the respective people(s)/individual(s) begin to meet each other face to face and learn about the other's inner world with depth (Tu, 1993; Watkins & Biggs, 1996). On the one hand, while Western countries often question China for its standardized exam system that oppresses the authenticity of learning (Nield, 2007; Ozturgut, 2011), they praise its success on Programme for International Student Assessment (PISA) and express greater interest in exploring the lessons behind such success in its general education system (Kermer, 2016; Tan, 2017). However, on the other hand Chinese educators, while proud of their high-achieving students' success, remain keenly interested in Western research outcomes and perceptions of higher education to enhance their educational sector (Dello-lacovo, 2009; Hayhoe & Pan, 1996).

2 Context of the Study

This study examines the experiences of collaborative learning for teachers in a cross-cultural context. It is situated in a longitudinal reciprocal learning partnership between Canada and China, led by Dr. Shijing Xu from the University of Windsor and Dr. Michael Connelly from the University of Toronto. In 2013, this partnership launched five research studies focusing on different subjects including mathematics education, language and cultural education, science education, teacher education, and general education. The partnership involves five universities in Canada and China. They are the University of Windsor, the University of Toronto, Northeast Normal University in China, Southwest China University, East China Normal University, and Beijing Foreign Studies University. As part of the project design, teacher pairs were formed and engaged in a series of different activities centering on teaching and learning within the Mathematics Education Team. The Mathematics Education Team recruited five kindergartens to grade 12 teacher pairs between 2014 and 2017, beginning with three pairs in 2014, one pair in 2015, and the last pair in 2017.

This study examines the partnership of the pair of middle school mathematics teachers who were paired in May 2017. The respective teachers come from the city of Toronto in the province of Ontario, Canada, and the city of Chengdu in the province of Sichuan, China. The symbolic learning activity for this pair of teachers was their collaboration in planning and teaching a mathematics lesson at the 5th Canada–China Reciprocal Learning Partnership Annual Conference, which showcased their partnership over the first year of collaboration.

2.1 Co-Teaching as a Model for Teacher Learning

This study engaged teachers in active teaching, assessing, and reflecting, with a primary focus on teachers' experiences of professional learning based on their teaching specific content. Darling-Hammond and Richardson (2009) state, in their review of teacher learning, that "the most useful professional development emphasizes active teaching, assessment, observation, and reflection rather than abstract discussions" (p. 47). Furthermore, professional development that focuses on student learning and helps teachers develop the pedagogical skills to teach specific kinds of content has strong positive effects on practice (Blank, de las Alas, & Smith, 2007; Wenglinsky, 2000).

We chose to engage teachers in a process of co-teaching because it is one of the effective ways that represent the outcomes of a successful partnership (e.g., Murphy & Martin, 2015; Sachs, Fisher, & Cannon, 2011). As Meier and Fisk (2016) recommend, a successful co-teaching experience requires the partnered teachers to first have a relationship where they could openly share about their background, philosophies in education, as well as their perceived weaknesses and strengths in teaching. Teachers develop trust and understanding by building relationships through sharing and discussions ranging from academic and professional concerns and beyond. From the beginning of their partnership, Annette and Ming have been comfortably exchanging information about themselves so that each partner was able to learn about the other's personal strengths and instructional practices as well as their personalities and personal preferences in daily lives. These engagements prepared the foundation for two teachers co-teaching activity.

A good co-teaching experience also requires teachers to plan together (Meier & Fisk, 2016). Taking advantage of social media tools allowed Annette and Ming to plan their lesson collaboratively. Planning the lesson together provided the opportunity for both teachers to "vary [their] teaching strategies" (p. 26). The collaboration between Annette and Ming reveals four key aspects: (1) How two teachers may differ in many ways with regard to planning and teaching, (2) what aspects of planning they emphasized, (3) what kinds of tasks teachers design, and (4) what primary skills that they intend to develop in students. Learning to share the lead is also recommended to achieve positive co-teaching between partners. In the process of co-planning and teaching a demo lesson, Annette and Ming chose different roles concerning when to take charge. Each perceived themselves as leaders for the class. Annette led the designing of the lesson plan, while Ming showed initiative in reaching out to the students and choosing the topic accordingly.

2.2 *The Teacher Pair*

The recruitment and selection of two teachers for this pairing happened coincidently rather than being purposefully planned. In this section, stories of how we encountered the teachers and why they volunteered to participate in this project will be presented. The partnership between this teacher pair represents the exchanges in mathematics teaching and learning at the high school level while the other teacher pairs were at the elementary level.

2.2.1 *The Chinese Teacher: Ming*

Ming is a mathematics teacher from a suburban middle school in Chengdu, Sichuan Province in China. She has ten years of experience of teaching grades 7–9 mathematics. She is a highly motivated teacher who continually seeks opportunities to learn and grow as a professional. Even though her school has limited resources, her principal supports and invests in all types of teacher learning.

In the Fall 2016, when Ming went to Germany to attend a Mathematics Education Conference, she met one of our research leaders of the Mathematics Education Team in the Partnership Project. Ming learned of the project from us and expressed her interest in participating in such a learning program that would partner herself with a mathematics teacher in Canada. To officially commence the partnership in May 2017, Ming was invited to come to Toronto for the 4th Annual Canada–China Partnership Conference and Annual General Meeting (AGM), creating the opportunity to meet with her partner, Annette. Ming had the opportunity to visit

Annette's school for one day, observed a couple of her lessons, and talked with some of the Canadian students during her stay in Toronto.

2.2.2 The Canadian Teacher: Annette

Annette is a Chinese–Canadian mathematics teacher in an urban public high school in Toronto. Annette was born in Hong Kong, China, and moved to Canada with her parents when she was a teenager. She has twenty years of experience in teaching the subject of mathematics in Canadian schools. She has currently been working in her school for over ten years. Annette is a highly motivated teacher who always brings creativity to her lesson planning and teaching while seeking to challenge herself with new learning experiences. As a relatively experienced teacher, she felt stuck in her career and hoped to expand her capabilities by learning with other teacher peers.

In early 2017, while attending a district-based professional development session at the Ontario Institute for Studies in Education of the University of Toronto, Annette met one of the research leaders of the Mathematics Education Team in the Canada–China Partnership Project. As Annette learned about the project, she expressed interest in participating in some ways. Given her Chinese background and language proficiencies in Mandarin Chinese, we consider her connection with the Chinese teacher will be different in comparison with other teacher pairs and valuable to this research project.

3 Data Resources

Data was collected from lesson-planning conversations between teachers (chat history from WeChat), the final lesson plan and related documentation, the videotaped lesson, post-lesson comments from on-site scholars, written reflections from teachers, and dialogical interviews conducted with each teacher participant to follow up with their written reflections.

3.1 Chat History on WeChat

WeChat is a tool for chatting, and it is one of the most popular social tools in China. Most people have the WeChat app installed on their smartphones. It allows making long-distance voice or video calls as long as both sides are connected to Wi-Fi. WeChat not only allows messaging but can also be used for sharing pictures and sending voices memos. WeChat also has built in translation abilities as well, so it helps the teachers between two countries to understand the other language without a physical translator. Since the beginning of the partnership, the research team has set up a chat group for each teacher pair and a facilitator, usually a research assistant, on WeChat.

After Ming returned to China after meeting Annette at the 2017 AGM in Toronto, Annette and Ming communicated through WeChat for one year. As the Mathematics Education Team planned for the 2018 Annual Conference and AGM in Changchun, we wanted to invite teacher pairs to co-teach as to showcase their learning partnership. In January 2018, we visited Annette's school to deliver the invitation. She accepted the invite and started communicating with Ming immediately. Between February and May 2018, they connected with each other online, discussing every detail of the lesson. All of these online communications, including messages in either voice or text, were collected as part of the data, together with any shared pictures and documents during the planning. Because of her Chinese background, Annette speaks Chinese and Ming speaks English. Therefore, they communicated mostly without a facilitator.

3.2 Written and Oral Reflections from Each Teacher

After the conference, each teacher was asked to write a reflection paper regarding their experiences of collaboration on the lesson, without any suggested format. Both teachers decided to reflect on the lesson using the Ten Dimensions of Mathematics Education (McDougall, 2004). The frameworks guided teachers to reflect on ten aspects of the classroom including program scope and planning, meeting individual needs, learning environment, student tasks, constructing knowledge, communicating with parents, manipulatives and technology, students' mathematical communication, assessment, and teacher's attitude and comfort with mathematics. Based on their written reflections, teachers were individually interviewed for one hour to elaborate on their main observations noted in the writing pieces.

4 Findings

4.1 *Collaboration: Designing the Lesson*

Because the lesson was going to be taught in a Chinese classroom, Ming thought that an original Canadian lesson would be more appealing to the audiences from Chinese backgrounds, so she encouraged Annette to be the lead teacher of the lesson to showcase an authentic Canadian lesson while Ming assisted her throughout the process. For the lesson design, Annette took critical thinking into consideration, as this had been the focus of their exchange in the last semester of partnership. Eventually, they decided to teach a lesson engaging students to solve problems in real-life contexts using linear function. Here is how the lesson was designed:

Annette created the teaching plan about two weeks before the teaching day. Both teachers discussed specific details to ensure that the design fit relevant curriculum, topic coverage, and the level of students being assigned at the conference. Ming translated the English plan into Chinese with help from the research team, and her colleagues, including the director of high school Research Education Department of Education and Science Institute of Sichuan Province. Therefore, Ming considered this lesson "an original Canadian Math class, to some extent, transferred to China classroom".

In order to understand the lesson and communicate it with the Chinese students, Ming asked questions regarding the plan to get clarification from Annette. Whenever she received feedback from the students, she updated to Annette, and whenever she had an idea, she communicated with Annette for advice. Ming showed photos of the classroom to Annette so that she got an idea of the setting.

On the night before the lesson in Changchun, China, Annette and Ming discussed the lesson one more time to make sure they were fully prepared. The next day, Annette and Ming arrived at the school early to communicate with students. They decided to limit the translation of the lesson to save time. Whenever necessary, Ming assisted on the clarification of instructions.

4.2 *The Lesson Goals and the Task*

For this lesson, Annette and Ming created real-life problems for students to solve using the linear equation model. The goals were to engage students in: (1) researching online and determining two cost structures for admission to a theme park (annual pass vs. day admission) "collaboratively"; (2) representing each cost structure using a linear relationship in the slope-intercept form: $y = mx + b$; (3) determining the intersection point of the two linear relationships and understanding its significance in determining which cost structure is a better deal under what conditions; and (4) considering other factors such as how many times they are able to visit the park to determine which cost structure is a better choice for them under their own circumstances (cost per visit vs. a high initial cost of an annual pass).

Annette first created the lesson plan using the 3-part lesson framework, and then they revised the lesson plan following expectations from teachers of the students who they will be teaching. The task is called "to deal or not to deal". They created a scenario for students to imagine that they will be buying tickets to Shanghai Disney for summer vacation. As many ticket types being offered to suit different family needs, students were asked to choose the best ticket type for themselves.

4.2.1 *The Actual Teaching Day*

On May 18, 2018, the lesson was co-taught and then discussed by assigned scholars in the project. The lesson and the commentary were videotaped. Teachers prepared bilingual PowerPoint slides to save time on the translation for the audiences. It was presented as a demo lesson in a lecture hall with state-of-the-art presentation technology. They had 54 students in that room. Random desks were placed next to the door, the teacher desk, and the window to accommodate all the students. There was not much spacing between the rows of desks. Annette reflected that the learning environment of the room is not set up for flexible grouping. The main lesson activity involved minimal direct instruction, a few pre-planned prompting questions along with a few impromptu scaffolding questions as guided facilitation, and lastly an extended explorative inquiry during which students could work independently or in pairs, with or without teacher assistance to arrive at their own individual responses to the posted question.

4.2.2 *Scholars' on-Site Commentary of the Lesson*

One Canadian professor graded the lesson as A++. He said that the lesson made connection to the students since everyone should have ridden some kind of ride in the past. It also made an emotional connection. Teachers circled around to help students. They made sure to let students of both genders shared their answers. The minds-on activity (the dragon boat/drop tower) also related to science like physics, integrating sciences and math. Both teachers felt a great sense of encouragement when one professor rated their lesson A++, “he commended how the lesson’s minds-on activity related to a common childhood experience that most of us have and associated the mathematical concept of height vs. time graph to scientific relationships”. Annette thought their minds-on activity acted effectively as a lesson hook to individual students.

Allowing students to justify their choices and later on giving them the opportunity to change their choices eased off the anxiety of the students being in the center of attention. They must have been under great pressure preparing for the demo lesson without really knowing what the lesson would be like. The minds-on activity put less emphasis on getting the right answer and more emphasis on understanding the relationship between height and time as well as experiencing how mathematical concepts can be applied in a real-life situation. It led well to the action part of the lesson.

5 Discussion

Collaboration in classroom teaching constitutes a key factor for success in schools. The ways that teachers collaborated revealed their visions of teaching and learning. Through collaboration, teachers learned about the differences between two cultures as well as reflected on their own practices from learning about the other culture. With enhanced understandings of the other culture, teachers were able to deal with differences and conflicts with their partners in the process of collaboration.

Teacher knowledge is central to understand teaching and learning as their firsthand experiences from the field are most authentic and valuable. As the subjects of learning, thoughts and opinions of those teacher cases were examined as the main source for knowledge construction in this study. Through guided reflections, teachers compared the two cultures and openly shared their perspectives on how culture related to teaching and learning. Based on the findings in the process of planning and teaching, we summarize differences in four aspects and further discuss how those differences influence and impact on teacher’s behaviors and perspectives of teaching and learning. In the following sections, we examine and compare two teachers in four aspects of their considerations in teaching.

5.1 *The Product or the Process of Learning*

According to our observations on teachers' collaborations during in co-planning, it seems that the Canadian teacher valued the process of learning as she expected students to make mistakes and go through struggles to make progress in learning, while her Chinese partner Ming focused more on making sure to avoid unexpected incidences for the whole class by perfecting each detail of the lesson.

Annette wrote in her reflection regarding the experience of co-planning the lesson and also pointed out this different emphasis that she observed between her Chinese partner and herself. Ming wanted every detail to be perfect during the planning stage and to include every question that might be posed. Annette said that her reaction would depend on the reaction of the students and the questions that they pose at that time.

We initially assumed that the main reason for Ming to care so much about the perfection of the lesson was merely for the showcase in front of the audiences at the conference. Nevertheless, in responding to Annette's comment, Ming argued that not only for the demo class, but she does the same for her own class in daily planning, because Ming believes that, if she spent more time on planning and being more careful on each detail, her students benefit more from the lesson. Ming said that when she was limited in time to prepare for a perfect lesson, she asked random questions and gave inaccurate answers during the class. In such cases, her students mostly got confused with what was being taught, and then Ming had to return to the lesson and spend more time on explaining to students.

Ming still thought it was necessary to spend more time on each detail because she simply did not want to waste time due to saving time initially. Especially for the demo class this time, which gave two teachers only 50 minutes in total to co-teach a whole lesson, Ming wanted to cherish every second and make the most of the time to achieve heuristic learning for all students possible.

Another aspect of this difference between product and process reveals that Annette considered the success of a lesson largely relates to the engagement of learners in the process of learning as well as students' abilities to apply the knowledge learned in solving real-life problems. Ming thought that a successful lesson must guarantee students' mastery of the knowledge in solving mathematics problems on paper. In the process of choosing the content to teach, initially Ming proposed to teach a review lesson of quizzes and test questions, which Annette did not agree to pursue, so they eventually decided to design a real-life application of mathematics knowledge.

The different emphasis on product or process of learning between Chinese and non-Chinese learners has been in discussion for decades, and it could root in the different evaluation systems in two countries. In China, the school preparation for the National Entrance Exam restrained teachers being creative in daily teaching because most of the time is used to implement numerous tests and quizzes throughout the school year. Ming said that because mathematics is one of the core subjects in secondary schools, it required priority on lessons exclusively for reviewing quizzes and test questions. Due to the quantity of content in the curriculum, teachers had to be very precise in each lesson. Even though Annette also experienced some extent of pressure from preparing students for the EQAO at certain grades in Ontario, she comparatively enjoyed much more freedom in her daily planning. Such freedom encouraged her to introduce applicable ways of mathematics learning, which resulted in the difference between herself and her Chinese partner in choosing content and the type of lesson to teach.

The pressure to take exams in China influenced the ways that teachers teach, and it also impacted on the manners of students in learning. Annette reflected on the students' behaviors that she observed in Chinese classrooms and in terms of this aspect, we discuss some related considerations and differences between two countries below.

5.2 Different Considerations About Student Behaviors

For this demo lesson, Annette did not put classroom management into consideration as she understood that the chosen students are among the most exceptional and highly performed in the city, but in daily teaching, Annette always prioritizes the needs and interests of her students. She explained that, in order to engage students and help them understand the importance of mathematics for their life, she always had to create tasks which closely relate to students. Meanwhile, because her students from Toronto schools usually required more assistance in understanding instructions during teaching, as many of them had special needs and not performing on the same grade level, Annette had to design different tasks to meet different levels of learning needs in one class. For Ming, student behavior was never part of the consideration in her planning, because her Chinese students at this stage are mostly disciplined to meet school standards, and on top of self-motivation, the assistance from parents made it easier for teachers to discipline students at school. Even for a regular classroom with different student performances, Ming rarely spends time on managing students during daily lessons.

One other factor that made the student management easier for Chinese teachers in middle school might be that, in China, one teacher usually teaches the same class of students for a minimum of three years. In Ming's case, she normally takes over a class of students from grade 7–9. She only spent time on classroom management for the beginning of the first year, and then it does not take much of her energy for the rest of middle school years.

The highly self-disciplined students in the demo class impressed Annette. As she hears about all the support that Ming has from the parents, Annette expressed her struggles in connecting to the parents in her school. Many students in Toronto came from immigrant families. In Annette's case, many parents are too busy struggling with adaptive issues living in a new country. She found it hard to reach parents whenever a conversation was needed because many parents are busy with work to meet their basic needs in life. As she understands the importance of frequent communications with parents, Annette always tries to be innovative in ways to connect with them. The different student behaviors in China stimulated further reflections on improving parents' engagement in student learning in Ontario schools for Annette.

Teachers' different considerations regarding student behavior are rooted in the local cultures where teachers come from. In a multicultural context where Annette teaches, she always takes the family background of each student into consideration, while for Ming, from a town where the traditions and cultures are consistent; students tend to behave similarly and took less energy to manage.

In this collaboration, the unfamiliar topics of special education and multiculturalism arise for Ming. She started to pay more attention to students with special needs in her school and reflect on some of the cultural-related issues in her teaching. Ming was proud of the behaviors in Chinese classrooms as Annette shares about her struggles. However, Ming started to question if students are well engaged in the learning while quietly sitting behind the desk and if students are thinking critically when they obediently nod and accept the knowledge delivered from the teacher.

5.3 National Policies of Education

In her reflection, Annette pointed out that she thought the educational policies in China had played a critical role in stimulating innovation in schools. When she attended the 4th Annual Conference in Toronto in 2017, her first impression was that China valued education as a means to economic growth. Her trip to Changchun in 2018 further validated her impression. When she walked into Northeast Normal University, she saw slogans on various lamp posts which says: "Better Teachers—A Stronger Country". From the slogan, Annette noted that China values education and teacher development. She thinks that educators in China have the mission to nurture their future generations who would create an even stronger China.

She was impressed with all the state-of-the-art technologies and facilities at the Chinese school she visited. Although the classrooms are a bit small with over 50 students sitting closely to one another, classrooms for "non-academic" subjects such as gym, music room, student kitchen, and art room are spacious. With education being a national policy, its goals and administrative policies are more unified.

Unlike in Canada, different provinces have different expectations and policies on education. This may allow provinces to develop policies that gear toward the needs of each province, but also turn educational processes into time lagging bureaucratic proceedings or political bargain chips during election times. When Annette attended a presentation delivered by a Chinese professor who proposed changes in the math curriculum and the Chinese student achievement evaluation system, she actually felt that he had the power and influences to make that happen because the bureaucratic process can be less time-consuming in China once you have support from the core political groups. However, she believes that, if a professor in Ontario is presenting systematic change for the Ontario educational system, she would feel intrigued, but she would not think any dramatic change would happen systematically soon. The reciprocal learning project allows Annette the opportunity to learn with other passionate educators outside Canada.

Ming agrees that, if scholars have good connections with officers, it is easier to make an impact on policy change and reform. However, it is not as easy as it seems in reality. Because this is sensitively related to politics, Ming did not want to go deeper in the discussion of such a complex topic.

6 Concluding Remarks

Effective professional learning for in-service teachers has sustainable effects on improving teaching and learning for students. Studying the experiences of teacher learning in cross-cultural contexts contributes to the research fields of teacher education and professional development while providing lessons and examples for improving the quality of learning programs for teachers. By using teacher voice as the main venue in understanding teaching and learning under cross-cultural context, it potentially contributed to the fields of teacher empowerment and knowledge mobilization for global education. In multicultural contexts like the Toronto schools, bringing together teachers' different "beliefs and values shaped by their own experiences of teaching, and being taught" (Chan, 2013) adds many colors and different meanings to the larger landscape of teacher learning while contributing to a more complete understanding of the phenomenon (Phillion, He, & Connelly, 2005).

Both teachers reflected on the importance of support from their partners in their learning and growth. To achieve a successful lesson, each teacher believed that the shared goals of openness and trust are the most critical components. Driven by the goal to improve student learning, teachers voluntarily participated in the learning activities of this longitudinal project. Over the years of partnership, these teachers built a trusted relationship affording a space of open sharing and knowledge construction. An inclusive community of learning is continually constructed within the project among the teachers, which extended to the researchers, as well as other stakeholders inside and outside of the schools. During co-planning, the teachers reached out to different parties to achieve effective learning. As the partnership was extended beyond each pair of teachers, teachers were able to learn from other teacher pairs.

Engaging teachers in collaborative activities such as co-teaching provided them with opportunities to actively learn and negotiate roles in teaching and learning processes. These teachers engaged in critical thinking and active inquiry through learning to teach with partners from a different culture. Each teacher took advantage of their knowledge and experience and took initiative to act and interact with their partner in the process. Chinese teachers, with the influence and support from their Canadian partners, became more courageous, taking risks and challenging themselves beyond their comfort zones. Meanwhile, Canadian teachers, with the cultural knowledge and new perspectives from China, began to reflect on different ways to improve their teaching across Canadian school contexts. Teachers consider themselves as global educators when they are actively involved in researching issues in global contexts.

References

Blank, R. K., de las Alas, N., & Smith, C. (2007). *Analysis of the quality of professional development programs for mathematics and science teachers*. Washington, DC: Council of Chief State School Officers.

Cai, J., & Knuth, E. (Eds.). (2011). *Early algebraization: A global dialogue from multiple perspectives*. Heidelberg, Germany: Springer. https://doi.org/10.1007/978-3-642-17735-4.

Chan, E. (2013). Teacher experiences of culture in the curriculum. In D. Flinders. & S. Thornton (Eds.), *The curriculum studies reader* (pp. 301–304). New York, NY: Routledge.

Cogburn, D. L., & Nanette, S. L. (2003). U.S.-Africa virtual collaboration in globalization studies: Success factors for complex, cross-national learning teams. *International Studies Perspectives, 4*, 34–51.

Darling-Hammond, L., & Richardson, N. (February, 2009). *Research Review/Teacher Learning: What Matters? How Teachers Learn, 66*(5), 46–53.

Dello-lacovo, B. (2009). Curriculum reform and 'quality education' in China: An overview. *International Journal of Educational Development, 29*, 241–249.

Dewey, J. (1926). America and the far East. Survey, 1 May, 1926, 188. Later published in: John Dewey, The Later works, 1925–1953, (1984). Carbondale: Southern Illinois University Press. Volume 2: 1925–1927, pp. 1173–1175.

Goodwin, A. L. (2020). Globalization, global mindsets and teacher education. *Action in Teacher Education*, *42*(1), 6–18. https://doi.org/10.1080/01626620.2019.1700848.

Hayhoe, R., & Pan, J. (Eds.). (1996). *East-west dialogue in knowledge and higher education*. Armonk, NY: M.E. Sharpe.

Hayhoe, R., & Pan, J. (Eds.). (2001). Knowledge across cultures: A contribution to dialogue among civilizations. Hong Kong: Comparative Education Research Centre, The University of Hong Kong.

Kermer, D. (2016). Why do Chinese students out-perform those from the West? Do approaches to learning contribute to the explanation? *Cogent Education*, *3*(1), 1248187.

McDougall, D. E. (2004). *School mathematics improvement leadership handbook* (p. 88). Toronto: Thomson Nelson.

Meier, B. S. & Fisk, J. S. (2016). Five ways to ensure a positive co-teaching experience for co-teachers and students. *Wisconsin English Journal, 58*(2), 22–33.

Moore, A. (2012). *Teaching and learning: Pedagogy, curriculum and learning*. London: Routledge-Falmer.
Murphy, C. & Martin, S. N. (2015). Coteaching in teacher education: research and practice. *Asia-Pacific Journal of Teacher Education, 43*(4), 277–280.
Nield, K. (2007). Understanding Chinese learner: A case study exploration of the Hong Kong Chinese Learner as a Rote or Strategic Learner. *Journal of Hospitality, Leisure, Sport & Tourism Education, 6*(1), 39–48.
Ozturgut, O. (2011). Standardized testing in the case of china and the lessons to be learned for the US. *Journal of International Education Research, 7*(2), 1–6.
Phillion, J., He, M. F., & Connelly, F. M. (2005). *Narrative and experience in multicultural education*. Thousand Oaks, CA: Sage.
Sachs, G. T., Fisher, T., & Cannon, J. (2011). Collaboration, mentoring and co-teaching in teacher education. *Journal of Teacher Education for Sustainability, 13*(2), 70–86.
Said, E. W. (1978). *Orientalism*. New York: Vintage Books.
Tan, C. (2017). Chinese responses to Shanghai's performance in PISA. *Comparative Education, 53*(2), 209–223.
Tu, W. (1993). *Way, learning, and politics: Essays on the Confucian intellectual*. Albany: State University of New York Press.
Watkins, D. A., & Biggs, J. B. (1996). T*he Chinese learner: Cultural, psychological and contextual influences*. Hong Kong: Comparative Education and Research Centre & Australian Council for Educational Research.
Wenglinsky, H. (2000). *How teaching matters: Bringing the classroom back into discussions of teacher quality*. Princeton, NY: Milken Family Foundation and Educational Testing Service.

PART III

Specific Content Topics in Mathematics

CHAPTER 9

Analysis of Students' Systematic Errors and Teaching Strategies for 3-Digit Multiplication

Yunpeng Ma, Shu Xie, Yanling Wang, and Douglas McDougall

1 Introduction

1.1 Teaching and Learning Plays a Decisive Role in Curriculum Reform

A classroom is a result of the combination of teaching and learning. It is wrong for educators and teachers to consider either teaching or learning

---

Y. Ma (✉) · S. Xie · Y. Wang
Faculty of Education, Northeast Normal University, Changchun, China
e-mail: mayp@nenu.edu.cn

Y. Wang
e-mail: Wangyl465@nenu.edu.cn

D. McDougall
Ontario Institute for Studies in Education, University of Toronto, Toronto, ON, Canada

S. C. Zhu et al. (eds.), *Reciprocal Learning for Cross-Cultural Mathematics Education*, Intercultural Reciprocal Learning in Chinese and Western Education,
https://doi.org/10.1007/978-3-030-56838-2_9

separately instead of bridging those two. At the center of pedagogy, learning should be the focus of curriculum reform. What is at the core of curriculum and educational reform is the change of learning conception, whose substance once grasped, will provide the powerful dynamic for curriculum reform (Huang, Huo, & Xu, 2014). Based on years of research, Huang and his team found that the key to learning is to let learners understand and engage, building their learning on understanding and setting it as the instructional goal.

The current trend in international mathematical education reform puts emphasis on *Teaching for understanding* (Hiebert & Carpenter, 1992), in which students should understand mathematics and instruction should center around students' understanding. *Teaching for understanding* encourages teachers to acquaint themselves with students' thinking, cognitive process, and its characteristics on learning and problem-solving and even conduct intellectual analysis in order to shed light on how students understand mathematics at the micro-level (Wildy & Wanace, 1992).

1.2 PCK Affects Teaching Effectiveness

Researchers have shown that teaching performance and its effectiveness are related to pedagogical content knowledge (PCK), which connects teachers and students by content knowledge (Gess-Newsome, 1999; Li, Ni, & Xiao, 2006; Ma, Zhao, Han, Song, & Zhao, 2008; Veal, Tippens, & Bell, 1999). Teachers with good PCK can flexibly deliver the meaning of content knowledge to students through appropriate teaching strategies and representation, thus boosting student learning (Gudmundsdottir & Schulman, 1987; Osborne & Wittrock, 1983).

Content knowledge is constant, but pedagogy knowledge is getting more complex, especially in aspects related to students. Its representative structure has transformed from simple linear formation to complex reticular one with students at its center, which makes knowledge of students boost the integration of PCK (Van Driel, Verloop, & de Vos, 1998). Only a deep understanding of student learning can promote quality instruction, so teachers should become reflective practitioners who ask the question of how students acquire knowledge and have the sense for constant instructional improvement (Huang et al. 2014).

Content knowledge includes teachers' understanding of specific content as well as students' misconception and errors, which calls for relevant representation strategy for instruction. A good representation involves the integration and internalization of content, learning, learners, and context and should be properly used in corresponding instructional scenarios (McDiarmid, Ball, & Anderson, 1989).

1.3 Student Errors Are Vital Instructional and Learning Resources

The earliest studies on mathematics errors were conducted in Germany, mostly from a psychological perspective, and the United States, based on behaviorism, which drew quite different conclusions due to political and cultural influences at that time. During the past century, having been explored from perspectives of social-culture, cognitive psychology, mathematics instruction, and student learning, research on mathematics error patterns and their cause analysis mostly concentrated in specific domains (addition, subtraction, variables, equation, fraction, decimal, and so on) as well as teachers' ability to predict errors and deliver error-based instruction.

The discussions about student error patterns have been carried about in specific domains and about certain content. On the one hand, error patterns can be achieved by interpreting students' learning difficulties (Holland, 1942; Osburn, 1946), such as operations of division, multiplication, subtraction, place value, quotient estimation, and inclusion or exclusion of zero. On the other hand, generalization of error patterns puts 3rd-grade arithmetic errors into four groups: wrong operation, obvious computation error, defective algorithm, and random response (Cox, 1975; Engelhardt, 1982; Radatz, 1980; Robert, 1968).

Cox (1975) classified addition algorithm errors into systematic, random, and careless errors. Research on algorithmic errors has been mostly carried out through the analysis of paper test in large samples. Some types were classified by student behavioral results, and others were based on performance during the learning process. Newman (1977) and Casey (1978) summarized these errors into four categories: comprehension, transformation, process skills, and carelessness.

Error analysis is a vital professional competence, which plays a key role in successful instruction. NCTM Principles and Standards for School Mathematics clearly indicate that effective mathematical instruction requires teachers to understand what students know and need to know and then assist them in learning. Teachers should be aware of student understanding difficulties of certain mathematics concepts as well as grasp methods of aiding students to overcome common errors. American psychologists Brown and Burton (1978) also agreed that modeling on students' internal errors is one of the greatest powers in teachers.

Detailed modeling of student knowledge, including errors, is the premise of effective error remediation. Some scholars in China attach great importance to teachers' thorough understanding of students' real thinking process and timely remediation of their errors (Zheng & Liang, 1998). Therefore, analysis of student errors will assist instructional research. Prediction of student errors in teaching plans and implementation of proper teaching strategies, opposed to the diagnosed errors in student learning, will contribute to better understanding of individual learning and prevention of student errors, thus aiding the process of instruction and learning, and boosting learning efficiency. Students will then not only acquire knowledge, but also realize how to learn (James & McCormick, 2009).

In summary, there is little research on teachers' PCK and teaching strategies in primary schools focused based on students' learning errors. The research has resulted in general conclusions and not on the substantive content knowledge. The student error research has investigated objective paper tests, but they provide little understanding of student thinking processes. The key research question is why are there such patterns in student errors and how do teachers react to the student errors?

It is possible, using qualitative research methods, to have deeper understanding of teachers' PCK by analyzing student erroneous thinking processes. In this chapter, we explore student and teacher understanding of mathematical errors.

1.4 Research Questions

There were three research questions posed to guide the research. The first question investigates common error types and patterns from the perspectives of external evaluators and teacher. Specifically, the question is "What are students' systematic error patterns, characteristics and reasons in 3-digit multiplication?". The second question investigates the teacher's pedagogical content knowledge as it relates to learning errors: "What are teachers' pedagogical content knowledge (PCK), recognition and analysis of learning errors on students' part?". The third question focuses on what decisions teachers make about the student's learning errors in the areas of awareness, diagnosis, analysis, and remediation of student learning errors.

Questions 3 is "What is the instructional decision-making process based on students learning errors?".

2 Theoretical Framework

Many scholars have tried to define pedagogical content knowledge (PCK) without reaching consensus (see Ball, Thames, & Phelps, [2008] and Shulman [1987], for example). PCK is considered to be the integration of student typical error knowledge, teaching strategies, and content knowledge in instructional practice. Integration of thorough content understanding, diversified instruction, cultural background, student prior knowledge, and experiences is the key to effective instruction, which aids students' acquisition of content knowledge in complex instructional scenario. Winsor (2003) used the three legs of the PCK Bench to represent content knowledge, knowledge of learners, and knowledge of pedagogy, in which the supportive relationship of the three legs maintained the bench's stability.

Based on comprehensive understanding of students and specific content knowledge, PCK is the knowledge that teachers use in specific instructional context to transform content knowledge to student understanding through the use of instructional strategies and representation. Its essence lies in integration of teaching and learning of specific content in specific context, new knowledge constructed both internally and externally (Xie, 2013). One of the key elements of PCK is knowledge of students, and it is placed at the center of PCK. Knowledge of representation and content knowledge are also important. Teachers' PCK should be researched with those three aspects, exploring its function in thinking and integrating aforementioned knowledge during instructional decision-making (see Appendix).

There are many types of errors students can make. It has been found that systematic errors are one of the main categories of errors. Cox (1975) has created an error classification model that identifies systematic error patterns and their causes. The repeated occurrence of systematic errors is the result of using the wrong algorithm or operation and caused by misconception or learning difficulties. In contrast, random errors are mainly due to impatience, lack of thinking, and obstructed memory retrieval. Those errors that fall short in offering causes or proofs are excluded from this research.

3 Research Methodology

3.1 Research Topic: 3-Digit Multiplication in Primary Mathematics

Mathematics Curriculum Standards for Compulsory Education (2011) divides primary mathematics into four parts with *Number and Algebra* as the main learning content and basis for other content. Through *Number and Algebra*, students understand formation and development of numbers, establishment and expansion of numerical relation, as well as learn about the concept of number. *Number and Algebra* will lay the foundation for the construction and learning of mathematics in other domains. The knowledge of *Number and Algebra* aims to cultivate students' number sense and operation ability. Therefore, the number operations is one of the core content in mathematics learning.

Armed with prior knowledge of multiplication, mental calculation of multiplication tables, multiplication of 1-digit with 2-digit and 3-digit, respectively, multiplication of 2-digit with 2-digit, the last concept taught is multiplication computation of 3-digit with 2-digit. Apart from the demands of right multiplication algorithm, understanding conceptions, and choice of reasonable procedure, students are also required to use this knowledge to solve real-life problems and offer explanation to the meaning of actual results as well as choose proper method of estimation in specific context. Since the selected content is the last phase of multiplication instruction, students' performance will reflect their overall grasp of whole number multiplication. Being more complex and diversified, the possible errors can be explained by any instructional phase during a three-year study of multiplication algorithm, thus making the analysis of those errors in this research more convincing.

3.2 Research Design

This study investigated teacher's perspectives of student's errors in Grade 4 classrooms while students explored and learned about 3-digit multiplication. This qualitative study followed two streams of data collection and analysis. The first stream provided insight into the teacher's performance and reflection. The second stream sought to understand student error patterns, causes of errors, and thinking processes. Figure 1 summarizes the research design processes.

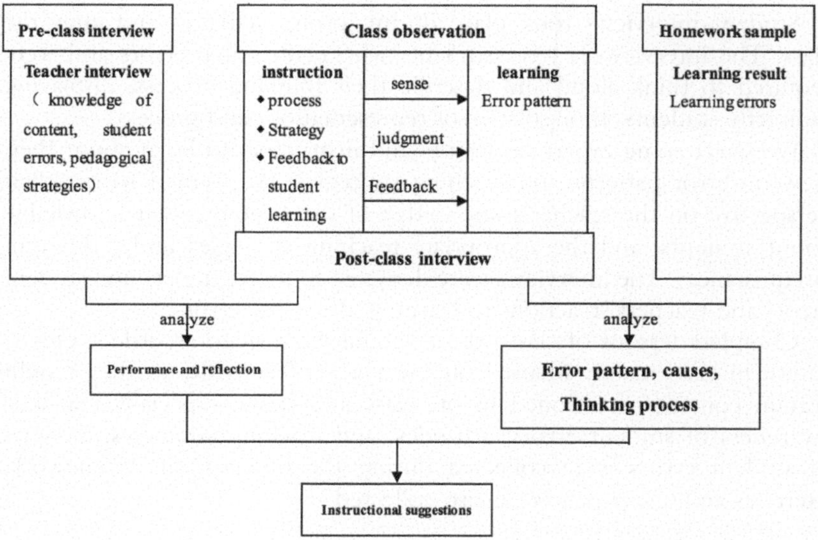

Fig. 1 Research workflow when collecting data, analyzing data, interpreting the analysis, and disseminating knowledge

3.3 *Participants: Teachers and Students*

There were two primary schools, E and F, chosen from two different cities in China. Three teachers, T-g, T-s, and T-m, were chosen from 4th Grade teachers in both schools. The teachers volunteered to participate in the study. The students were the Grade 4 students in these teachers' classroom. All participants consented to the study.

3.4 *Methods of Collecting Data*

Teachers were interviewed to gain insight into the teacher's understanding of mathematics content, their knowledge of students, common errors and misconceptions in mathematics, and the challenging elements of their teaching. Questions were also posed about the reasons why students make certain errors and the teaching strategies they employ when helping students who make certain types of errors. These pre- and post-class interviews were designed to investigate the similarities and differences between the planned lesson and the actual class.

Student interviews took place during group activities and after the class. The interviewees were students who made some errors and were required to think aloud and describe their thinking process, displaying indirectly students' thoughts, error representation, and process.

We select some expert professors and instructors to learn about their view on error patterns and teaching strategies. We wanted to get their perspective on the teacher's knowledge of the content, their knowledge about students, and the appropriate teaching strategies under different circumstances. The interviews were designed to better understand student errors and teachers' reactions to learning about the errors.

Complete lessons of classroom teaching were video-recorded. Observation in class mainly focused on the teachers' activities as they taught certain contents mentioned in pre-class interviews, especially teachers' awareness of student errors, attitudes, and teaching strategies. Samples of student errors were collected during the observation. Homework, exercises and test papers were also collected.

3.5 Data Analysis

The data that were collected included student homework, student think-alouds, classroom recordings, teacher interviews, student exercises, and test papers. These data were analyzed by categorizing the error type and creating proportional distributions. The various types of errors were counted. We asked two external evaluators (subject expert E-m and primary mathematics researcher E-w) to review the interviews with the teachers and the classroom recordings. The experts used the PCK rubric (Xie, 2013, revised from Park & Oliver, 2008) to provide evidence of the teacher's PCK. They also reviewed the data analysis and gave some feedback. The evaluators' scoring coefficient toward three teachers were $rs = 0.661$, $p < 0.01$; $rs = 0.732$, $p < 0.01$; and $rs = 1$, which resulted in relatively high consistency implies the results are reliable.

4 Results and Discussion

4.1 Students' Error Types

The teachers and experts identified a number of errors that were made by the students. They calculated the error frequency among 1595 computation items of 3- and 2-digit multiplication questions. The total error frequency is 85, accounting for 5.33% of whole sample. The random error frequency is 5, including copying the wrong topic and the wrong numbers, accounting for 5.88% of all error types. The largest group of errors are systematic errors, which account for 94.12% of the student errors. As we are focusing on systematic errors, we did not include careless errors or random errors in our analysis. Typical error types and related details are as described in the following sections.

4.1.1 *Multiplication Misconception*

The key point of this content is to let students understand multiplication procedure of 3-digit and 2-digit as shown in Fig. 2a. During two periods of instruction and exercises, error frequency caused by procedural misconception is 13, accounting for 15.29% of all error types. This implies that students failed to grasp the multiplication procedure of "Cross-multiply the multiplicand by the ones digit and the tens digit of the multiplier, then add the two partial products together" (interview, E-m).

Some of the students understood the procedure, but failed to clarify the meaning of ***multiplied the multiplicand by tens digit of the multiplier***, thus resulting in misconception of place value. Error frequency is 7, accounting for 8.24% of all error types. Problems are as shown in Fig. 2b.

4.1.2 *Computational Error*

Error frequency of this type is 62, accounting for the 72.94% of all error types. Multiplication error frequency (Fig. 2c) is 33, accounting for 38.82%; addition error frequency (Fig. 2d) is 15, accounting for 17.65% (shown as follows). Multiplication errors happened in all three classes, reflecting students' loose grasp of multiplication tables.

The error frequency of carried number is 10, accounting for 11.76% of all error types (Fig. 2e); the frequency of forgotten carried number is 6, accounting for 7.06%; the frequency of carrying more digit forward (Fig. 2f) is 2, accounting for 2.35%; the frequency of wrong carried number (Fig. 2g) is 2, accounting for 2.35%, which shows students' lack of proficiency in carrying principles.

In addition, the error frequency of "0" (Fig. 2h) is 4, accounting for 4.71%. It is worth noting that students with "0" error are all in T-m's class, which means the teacher had failed to help students understand the meaning of "0" in tens digit and in ones digit, resulting in students' computation errors.

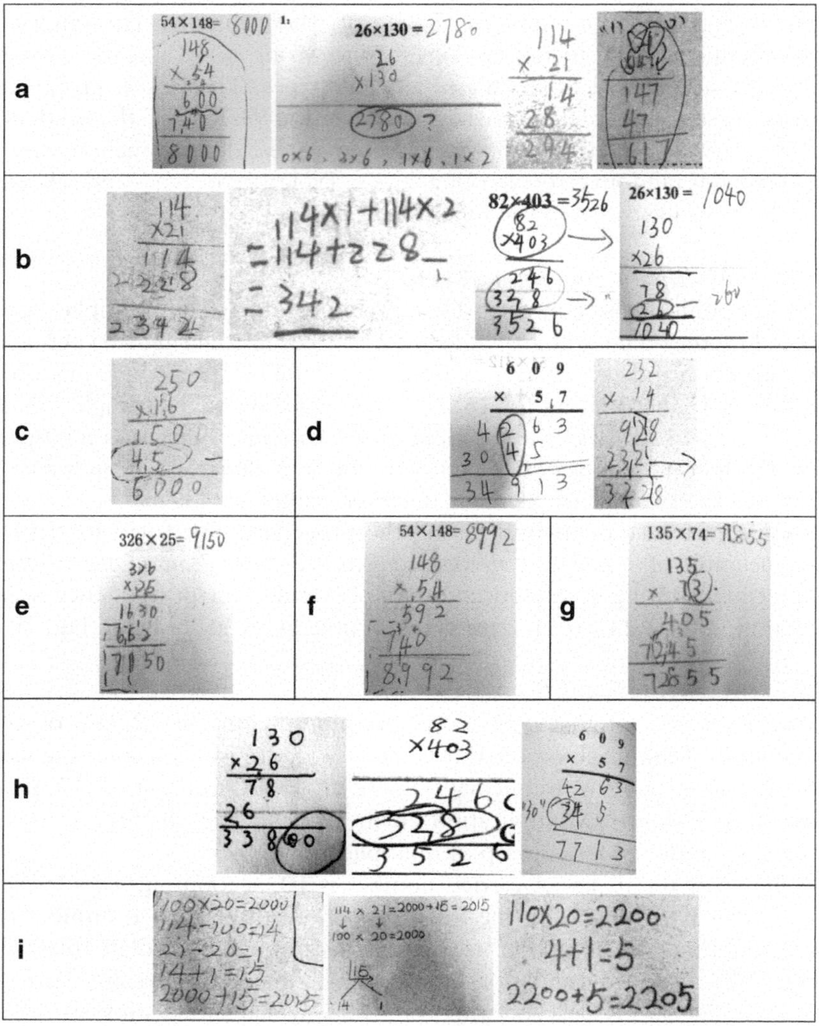

Fig. 2 Different types of student's errors when learning 3-digit by 2-digit multiplication

4.1.3 Procedural Error

In multiplication algorithm of "114 × 21", students used multiplication decomposition to get the results, but failed to understand the place value of decomposed numbers, which led to faulty results (Fig. 2i). The error frequency of this type is 5, accounting for 5.88%.

4.1.4 Summary

We learned that error distribution varied in all classes. The major mistakes in T-g's class were procedural misconception and faulty decomposition. In T-s' class, students made mistakes during computational process. In T-m's class, students had a high proportion of procedural misconception and calculating mistakes. Teacher T-g did not offer good scaffolding experiences during procedural instruction to help bridge decomposition and vertical multiplication. Teacher T-s put more emphasis on the conception than the procedure, while teacher T-m focused on difficult and challenging exercises without students' thorough understanding of conception and familiarization with the procedure, leading to higher error proportion.

4.2 Teachers' Understanding of Student's Error and Awareness Actions

4.2.1 Teachers Presuppositions and Understanding of Students' Errors

The three teachers indicated that they have a good understanding of student errors. The teachers are at a "proficient" level in being able to evaluate the learning environment. They were at the "proficient" level in understanding students' prior knowledge (such as the multiplication table, multiplication of 2-digit with 2-digit, and multiplication of 3-digit with 1-digit) and basic mathematics competence (various computational methods such as vertical algorithm and factor split). They could analyze students' errors in this learning content. T-m realized that many students' errors were due to erroneous procedure, especially during the error-prone process in columns alignment.

As for the required computational diversification, T-m presupposed that the majority of students would use the familiar method of factor split instead of "the rarely used vertical algorithm which may not occur to students". T-m was not able to provide the researchers with enough examples for students' recurring errors. He incorrectly attributed most of the errors to the student's failure to use the correct procedure. In contrast, T-g and T-s gave more comprehensive examples and analyzed specific mistakes. For example, T-g pointed out that the students "forgot renaming", had "unclear carried digits", "misaligned columns", and "vague memory of multiplication tables".

T-g also pointed out the "in vertical algorithm, students like to put 23 above, and 408 below; they don't understand the '0' in tens column means 0 ten, thus making calculation mistakes". T-s pointed out that students were confused with the four arithmetic operations and were sometimes confused with the multiplication procedure. She also pointed out that some of the students had a vague memory of multiplication tables.

As compared to the above two teachers, T-s gave a more coherent, comprehensive classification and attribution analysis (E-m, E-w). However, the teachers did not predict some of the students' errors, but during the interview, they reflected and analyzed as follows (Table 1).

Table 1 Students' error pattern beyond teachers' presupposition and causal analysis

Teacher	*Errors*	*Performance & reasons*
T-g	54 × 312	Misaligned columns in vertical algorithm, especially with 2-digit above, and 3-digit beneath, due to students' failure in grasping vertical algorithm
	114×21=2015 ↓ ↓ ↑ 100×20=2000 ↓ ↓ ↑ 14 +1=15	Split error due to misunderstood procedure
T-s	114×21 =114×（20+1） =114×21	Students have never learned about the splitting principle of multiplication, therefore meeting troubles when splitting numbers. They are also not good at making calculations of long formulas
T-m	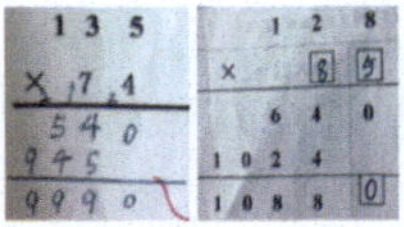	Never expected students to be unable to solving these problems

4.2.2 *Teachers' Inadequate Awareness of Students' Error Patterns*

In the study, it was observed the teachers could predict and realize most errors made by students, a fact which made them aware during the teaching process. It was further established that they could not take care of all of the students' errors due to time constraint. In so doing, the teachers simply attributed the errors to sloppy calculating, but gave no detailed strategies that could help students to analyze which part resulted in the calculation error. Instead, they gave students simple suggestions as "wrong, try again" or "be more careful". There were a number of errors that seemed to continue in T-m's class. It showed that the teacher was unaware of the errors and how to correct them.

In the instructional strategies based on the specific issue, teachers either put too much emphasis on the unity and clarity of computation or gave inaccurate professional opinions. As shown in Fig. 3, T-s judged students' answer as inaccurate because they "did not answer in accordance with the

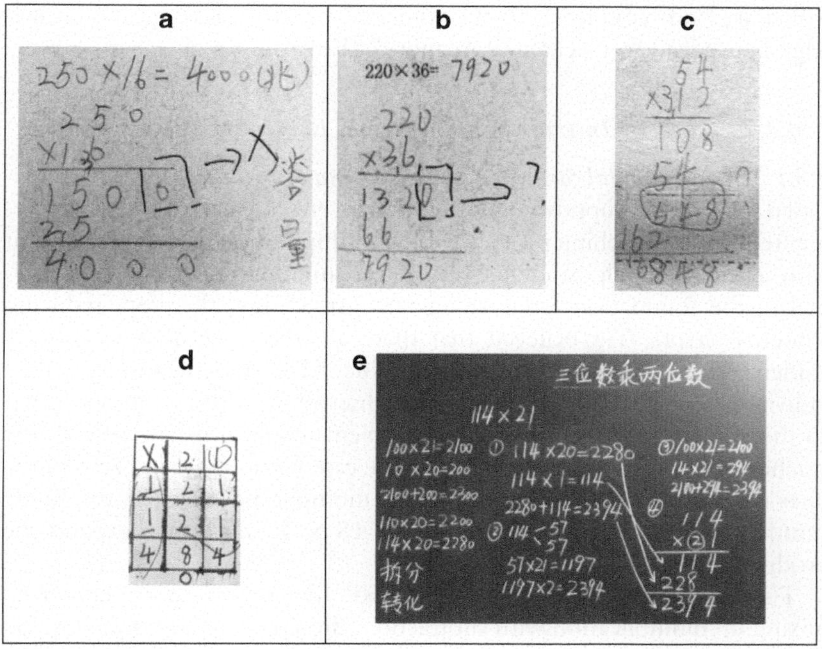

Fig. 3 Teacher corrections and awareness of student errors when teaching

procedure hence failed to reflect the mathematical brevity" (T-s). In fact, Fig. 3a and b shows that, since students could not calculate proficiently, they reminded themselves by "supplementing 0". In Fig. 3c, students converted "3-digit multiplying 2-digit" into their prior knowledge of "ones and tens columns of multiplicand multiplied by multiplier" and "hundreds column of multiplicand multiplied by multiplier" and then added the partial products together.

During interviews, the teachers mentioned that:

> this is generated by knowledge base and learning habits. Her knowledge base is not solid, related to her prior knowledge and cognition. Her errors today exposed the previous misunderstanding of multiplication procedure ...I do not think the problem can be solely attributed to attitudes and habits, but also can be explained by her mathematical understanding......there are rules for vertical algorithm of multiplication.

But teachers blindly emphasized the "universal", "common" methods, which are "not conducive to the students' creative thinking and knowledge hence lead to mechanical memory" (E-m).

4.3 Teachers' Instructional Strategies Related to Student Errors

4.3.1 Instructional Strategies on Improving Student Errors

Based on the presupposition of students' errors, teachers have implemented specific teaching plans. By using multiplication exercises of 2-digit with 2-digit to help students review the multiplication procedure, T-m laid the foundation for learning new knowledge, therefore reducing students' errors. T-g believed that there are three methods to remedy student errors when it comes to place value. "The students can be guided individually or through group and class discussion, with the questions left to the students themselves". If the problems could not be resolved, the teacher would then offer guidance. In the case where there is a zero in the ones column of either the multiplier or multiplicand, teachers can guide "students to put 0 aside and deal with 48×17 first, then expand the product tenfold".

T-s has "not yet found a good method" for preventing students from mixing up multiplication with subtraction, division, and addition. T-s also believes that students who make mistakes performing multiplication or addition when following the correct procedure simply need to practice more by doing "more exercises". For procedural errors, T-s stated:

> the students need to look carefully at each stage of the multiplication algorithm. Students should be able to notice if they have missed a step in their calculation. Teachers guide this discovery by circling specific numbers, using arrows to indicate where steps may have been missed.

4.3.2 Using Appropriate Instructional Strategies

Teachers were able to provide appropriate teaching activities based on the knowledge of students and their understanding of the content. The main teaching activity was to provide examples from the textbook and to use a variety of methods to calculate the "114 × 21" and then share their methods within groups and report one of their methods. During the process of independent calculating and group discussions, the teacher posed questions to students after they observed the students' calculations. For some common and typical algorithms (including the use of an erroneous algorithm), teachers could display them in the class and let the students to verbalize the process of calculating and explain their reasons.

For those typical errors, the teacher asked the students to challenge each other by asking questions until the misconception was clarified. If students made mistakes in calculating "54 × 312", T-s would ask "How is the vertical algorithm arranged? What is put above? What is put beneath? What is multiplied first? What is multiplied later?" When the student replied, "multiply 312 by 5", the teacher raised his voice and repeated "Is it 5?" until the student answered "312 multiplied by 50", and then teachers made a conclusion by stressing important points and precautions.

As shown in Fig. 3d, students explored a variety of methods to explore multiplication of 3-digit with 2-digit, but they made mistakes by tabling method. T-m identified it on the exhibition stand based on calculation principle. Teachers asked the students: "First step, which number do we use to multiply? Where should we put it? Second step, what number should we use to multiply? Where do we put it?" The teacher showed that 2 multiplied by 114 actually means 20 multiplied by 114 so the result is 2280. When explaining the zeros, the student should enter the zeros on the grid (see Fig. 3d).

Based on the evaluation of experts, the teaching strategy proficiency of the teachers is on the level of "Excellent", "Good", and "Pass". T-s and T-m made proficient performance in teaching content, using strategies for instructional key points, and challenging students' misconceptions, learning mistakes, and learning difficulties. The three teachers often use these expressions to explore students "Who understood?", "How do you think about it?" T-s mentioned in the interview that:

> It is not advised to tell students how to solve the problem directly. Instead, teachers should encourage students to solve the problem by observing, questioning and discussing, hence find problem-solving ideas by themselves. When students fail to reach consensus, I would not intervene to guide unless they cannot completely handle it. The logic enables students to have a deeper understanding of the mathematical procedure. As for such questions, I would deal with them more carefully.

Based on students' feedback, teaching strategies and their adaption made by T-m were on the "Pass" level; T-g's teaching proficiency in the abovementioned three aspects fell somewhere between "Good" and "Pass".

Data show that all three teachers should try to improve their post-class reflection. For instance, T-g was not specific enough in the analysis of students and teaching strategies for improvement. Instead, he thought that many students would be able to improve solely by practicing more.

5 Conclusion

The students displayed three types of systematic errors when multiplying 3-digit with 2-digit numbers together. The students displayed three typical types of systematic errors: computational error, misconception of multiplication, and erroneous procedure, accounting for 72.94, 15.29, and 5.88%, respectively. Teachers need to pay close attention to computational errors that are made by their students. In many cases, students need to be more fluent with multiplication facts while other students need support in learning place value during the use of partial products. There are also errors in lining up the columns to better understand the use of place value.

5.1 Teachers Are Able to Predict Students' Errors and Ability of Rules Exploration and Application Should Be Further Improved

The research shows that teachers had some knowledge of students' errors types, could analyze the causes of errors, and plan appropriate teaching strategies. Teachers need to consciously discover and identify student's errors in teaching and further analyze their causes. Among all of the teaching strategies, teachers preferred to use guiding questions with students, as a means of helping them to clarify errors and get the right answers. However, facing students' errors, the teachers were unable to realize the instructional value of errors and analyze the types of errors, proportion, and related instructional adjustment consciously. For example, all of the three classes had some types of errors, but they did not collect or analyze the main existing problems during instructional process.

Dealing with the issue, the teachers paid more attention to individuals rather than students with the same type of errors. In T-m's class, the same errors occurred through multiple classes. As in the first lesson, he did not give enough attention to students who easily made mistakes and correct their errors. In the second lesson, he did not identify the students making mistakes and give them differentiated instructions. T-g teacher should emphasize algorithms teaching, while T-s teacher should emphasize not only procedure teaching but strengthen procedure enforcement and its exercises.

5.2 Teachers' Understanding the Subject Matter Knowledge and Students' Errors Have an Impact on Their Instructional Implementation

According to the experts' evaluation, the three teachers' subject knowledge is in the "skilled" level. During interviews, teachers clearly knew that 3-digit multiply with 2-digit was the highest requirement for mathematics multiplication in elementary school. The core of multiplication is computing power, but its nature is to grasp multiplication procedure and carry out the calculation based on the procedure. In the actual classroom teaching, the specific teaching strategies of the three teachers vary. The results of expert evaluation show that T-s' proficiency in grasping instructionally difficult and key points was on the "Excellent" level, T-m teacher on "Good" level, and T-g teacher on the "Pass" level. For example, T-g had some bias in understanding estimation (E-m, E-w), which took six minutes during class time as compared to other teachers, who spent 2 minutes 20 seconds.

The teachers' estimation did not play a role in verification when calculating, resulting in students' lack of enough time to consolidate and practice new knowledge. The teacher also failed to find students' errors and offering related instruction. T-s and T-m understood the function of estimation in the lesson and were able to list the estimation results on the blackboard according to the order that allowed students to feel the range of estimation. T-m, especially, reflected the estimation from the beginning to the end to let students know its function.

When T-m asked students to elaborate on the exploring process of multiplication algorithm of 114×21 through diversified methods, aligning splitting steps with vertical algorithm (See Fig. 3f). T-m also expressed a strong intention of training by posing difficult questions, which reflects that the teacher paid more attention to hard and challenging questions (E-m).

The teachers tended to design and implement teaching plans based on personal understanding, ignoring previous strictly followed instructional goals. Therefore, teachers with higher abilities can find core content in lesson quickly. Unfortunately, this also means that teachers' behavior could also divert student attention away from important learning content.

5.3 *Influence of Textbook Arrangement on Teaching and Learning*

The teaching material is the medium between teacher's teaching and student's learning. It plays a very important guiding function, but the specific design or content of the textbook could mislead teachers and students. For example, the teacher will allocate time to a topic, such as the design of estimation, depending on the importance placed on the topic in the textbook. A number of misconceptions can arise by the use of certain materials. The use of graph paper could affect student understanding of place value, thus leading to the student's application of the graphing method and misconception of place value.

Clearly, textbook arrangement affects teachers' understanding of the curriculum and the teaching implementation. After the typical examples of the teaching materials are presented, if teacher does not help students analyze their deep mathematical meaning, it will impair students' understanding of specific contents and give rise to students' errors.

6 Implications

6.1 *Student's Errors: A New Perspective to Study Teachers' PCK*

At present, there is an abundance of international research on pedagogical content knowledge (Cankoy, 2010; Cochran, DeRuiter, & King, 1993; Halim & Meerah, 2002) involving case studies to directly explore the different types of teachers, which include novice teachers, experienced teachers, expert teachers, and teachers. However, there are a few cases that explore teachers' pedagogical knowledge in terms of knowledge of the student. Knowledge of students is at the core of PCK-integrated knowledge, which affects teaching strategies. Students' errors are relatively important in knowledge of students, which can be affected by prior knowledge, misconceptions, and learning difficulties. Teachers' understanding of students' errors can reflect the degree of their grasp of the nature of academic knowledge and knowledge of students, which can be used to judge the appropriateness of teachers' related instructional strategies.

6.2 *Curriculum Resources to Promote Teaching Research and Practice*

Students' errors cannot be explained simply as "situational accidents". In fact, many errors cannot be simply attributed to unawareness, carelessness, and other reasons. Some errors are recurring, regularly, with a certain continuity (unless teachers can offer proper guidance), which can be attributed to specific difficult experiences in learning or some external interference (Radatz, 1980). Therefore, analyzing the types of errors and causes behind them is crucial for exploration of students' errors to their teaching value. Through analysis of students' errors, exploring their features, patterns, and causes, we find that evidence-based teaching plan and differentiated instruction can be implemented in correcting these errors. The curriculum design may also need to be modified to include commons errors to teach and specific exercises to reinforce learning.

While teaching experience is valuable, there may be more efficient ways of gaining understanding of student knowledge and thinking. Through collective lesson planning, teaching seminars, learning exchanges, and other activities, we can summarize from these aspects and help each other enrich teaching knowledge, meanwhile, make a full preparation for theoretical analysis and strategic preparation. In addition, teachers should actively reflect post teaching, as well as adopt improving strategies in order to enhance teaching quality and help students learn better.

6.3 A Way to Help Students' Mathematical Communication and Thinking

With the continuous development of education and reform, requirements for students shift from mastery of knowledge, skills, and correct operation to students' awareness of one's own learning and the cultivation of thinking ability to enhance mathematical communication is possible. By improving communication between teachers and students through proper usage of errors and misconceptions in teaching, teachers are able to encourage students to verbalize their opinions and thoughts. In-depth analysis of students' learning errors will help teachers modify their PCK to provide students with timely correction of those errors which will help them develop a more stable foundation and sound knowledge system.

Appendix 1

See Table 2

Table 2 Analysis of PCK (adapted from Park & Oliver, 2008)

	Core knowledge dimensions	*Specific content*	*Grade level (1–4)*			
			limited	*pass*	*good*	*excellent*
Plan	Pedagogical content	Understanding of subject matter knowledge (including the understanding of its nature)				
		Grasp of important and difficult teaching points				
	Students' knowledge	Understanding of students' prior knowledge				
		Understanding of students' learning difficulties, misconceptions, & learning mistakes				
	Knowledge of instructional strategy & representation	Instructional strategies based on prior knowledge				
		Instructional strategies proper to learning difficulties and misconceptions				
Implement	Knowledge on students' understanding	Exploration on students' understanding				

(continued)

Table 2 (continued)

	Core knowledge dimensions	*Specific content*	*Grade level (1–4)*			
			limited	*pass*	*good*	*excellent*
		Rationale for instructional strategies & representations in connection with student understanding				
	Knowledge of instructional strategy and representation	Strategies based on content, important and difficult teaching points				
		Challenge on students' misconceptions, learning error and difficulty				
		Instructional strategies and debugging based on students' feedbacks				
Reflection	Knowledge on students' understanding	Difference between students' understanding and their real level				
		Students' misconception, learning difficulty and error				
	Knowledge of instructional strategy & representation	Rationality of implemented instructional strategy & representation				
		Plan to improve instructional strategy & representation				

References

Ball, D. L., Thames, M. H., & Phelps, G. (2008). Content knowledge for teaching: What makes it special? *Journal of Teacher Education, 59*(5), 389–407.

Brown, J. S., & Burton, R. R. (1978). Diagnostic models for procedural bugs in basic Mathematics. *Cognitive Science, 2,* 155–192.

Cankoy, O. (2010). Mathematics teachers' topic-specific pedagogical content knowledge in the context of teaching a^0, 0! and a/0. *Educational Sciences: Theory and Practice, 10*(2), 749–769.

Casey, D. P. (1978). Failing students: A strategy of error analysis. In P. Costello (Ed.), *Aspects of motivation* (pp. 295–306). Melbourne: Mathematical Association of Victoria.

Cochran, K. F., DeRuiter, J. A., & King, R. A. (1993). Pedagogical content knowing: An integrative model for teacher preparation. *Journal of Teacher Education, 44*(4), 263–272.

Cox, L. S. (1975). Systematic errors in the four vertical algorithms in normal and handicapped populations. *Journal for Research in Mathematics Education, 6,* 202–220.

Engelhardt, J. M. (1982). Using computational errors in diagnostic teaching. *Arithmetic Teacher, 29*(8), 16–19.

Gess-Newsome, J. (1999). Pedagogical content knowledge: An introduction and orientation. In *Examining pedagogical content knowledge* (pp. 3–17). The Netherlands: Springer.

Gudmundsdottir, S., & Shulman, L. (1987). Pedagogical content knowledge in social studies. *Scandinavian Journal of Educational Research, 31*(2), 59–70.

Halim, L., & Meerah, S. M. (2002). Science trainee teachers' pedagogical content knowledge and its influence on physics teaching. *Research in Science & Technological Education, 20*(2), 215–225.

Hiebert, J., & Carpenter, T. P. (1992). Learning and teaching with understanding. In D. A. Grouws (Ed.), *Handbook of research on mathematics teaching and learning: A project of the national council of teachers of Mathematics* (pp. 65–97). New York: Macmillan.

Holland, H. (1942). Difficulties involved in long division and some suggestions for teaching the process. *The Elementary School Journal, 42,* 585–596.

Huang, X., Huo B., & Xu, H. (2014). *Modern theories in learning and teaching.* Beijing: People Education Press.

James, M., & McCormick, R. (2009). Teachers learning how to learn. *Teaching and Teacher Education, 25*(7), 973–982.

Li, Q., Ni, Y., & Xiao, N. (2006). Subject teaching knowledge of elementary mathematics teachers: Research on performance characteristics and their relationships. *Journal of Education, 4,* 56–58.

Ma, Y., Zhao, D., Han, J., Song, X., & Zhao, X. (2008). Investigation of the professional knowledge of middle school teachers. *Journal of Northeast Normal University, 6,* 57–64.

McDiarmid, G. W., Ball, D. L., & Anderson, C. W. (1989). Why staying one chapter ahead doesn't really work: Subject Specific pedagogy. In M. C. Reynolds (Ed.), *Knowledge base for the beginning teacher* (pp. 193–205). New York: Pergamon.

Newman, M. A. (1977). An analysis of sixth-grade pupils' errors on written mathematical tasks. In M. A. Clements & J. Foyster (Eds.), *Research in Mathematics education in Australia, 1977* (Vol. 2, pp. 269–287). Melbourne: Swinburne College Press.

Osburn, W. J. (1946). Levels of difficulty in long division. *The Elementary School Journal, 46,* 441–447.

Osborne, R. J., & Wittrock, M. C. (1983). Learning science: A generative process. *Science Education, 67*(4), 489–508.

Park, S., & Oliver, J. S. (2008). National Board Certification (NBC) as a catalyst for teachers' learning about teaching: The effects of the NBC process on candidate teachers' PCK development. *Journal of Research in Science Teaching, 45*(7), 812–834.

Radatz, H. (1980). Students' errors in the mathematical learning process: A survey*. *For the Learning of Mathematics, 1*(1), 16–20.

Robert, G. H. (1968). The failure strategies of third grade arithmetic pupils. *Arithmetic Teacher, 15*(5), 442–446.

Shulman, L. S. (1987). Knowledge and teaching: Foundations of the new reform. *Harvard Education Review, 1,* 1–22.

Van Driel, J. H., Verloop, N., & de Vos, W. (1998). Developing science teachers' pedagogical content knowledge. *Journal of Research in Science Teaching, 35*(6), 673–695.

Veal, W. R., Tippins, D. J., & Bell, J. (1999). *The evolution of pedagogical content knowledge in prospective secondary physics teachers.* ERIC. ED443719.41.

Wildy, H., & Wallace, J. (1992). Understanding teaching or teaching for understanding. *American Educational Research Journal, 29*(1), 17–28.

Winsor, M. S. (2003). *Preservice Mathematics teachers' knowledge of functions and its effect on lesson planning at the secondary level.* Doctoral dissertation, University of Iowa, Iowa, IA.

Xie, S. (2013). *Structure and characteristics analysis on primary Mathematics teachers' pedagogical content knowledge.* Doctoral dissertation, Northeast Normal University, Changchun.

Zheng, Y., & Liang, G. (1998). *Cognitive science: Constructivism and mathematics learning.* Shanghai: Shanghai Educational Press.

CHAPTER 10

Utilizing Multiple Methods in Mathematics Problem Solving: Contrasts and Commonalities Between Two Canadian and Chinese Elementary Schools

Anthony N. Ezeife, Aihui Peng, and Bo Yu

1 Theoretical Background

Studies on comparisons of the Eastern and Western paradigms or systems of education abound in the research literature. For instance, in her contribution to the research discourse, Zappia (2018) quoted the ancient

A. N. Ezeife (✉)
University of Windsor, Windsor, ON, Canada
e-mail: aezeife@uwindsor.ca

A. Peng · B. Yu
Southwest University, Chongqing, China
e-mail: 23166380@qq.com

B. Yu
e-mail: 1354701002@qq.com

S. C. Zhu et al. (eds.), *Reciprocal Learning for Cross-Cultural Mathematics Education*, Intercultural Reciprocal Learning in Chinese and Western Education,
https://doi.org/10.1007/978-3-030-56838-2_10

Chinese philosopher, Confucius, who said that: "Education breeds confidence. Confidence breeds hope. Hope breeds peace." Interpreting and aligning this philosophical stance to educational policies and development has resulted to the situation whereby 'Eastern' countries largely root their educational thoughts and programs on the Chinese/Confucian Heritage Culture (CHC) tradition while, in 'Western' nations, education is rooted in the "Classical Greek view advanced by Socrates…" (p. 1). Expatiating on this contrast between Eastern (Asian) countries and their Western counterparts (of European cultural heritage), Zappia (2018) noted that Eastern societies focus more on the collective needs of the group, as opposed to the situation in the West, which is individualistic in orientation.

Classroom activities and interactions in the field of mathematics education have been of immense interest to researchers and educators in recent decades. Thus, many comparative studies in this area of research engagement have been undertaken in several countries with distinct or differing cultures. Some of these comparative studies in the East-West systems of education have highlighted obvious distinctions between the "Eastern" and "Western" educational practices. For instance, Kaiser and Blömeke (2013) noted that, in the field of mathematics education, the possession or acquisition of essential basic mathematical knowledge is the focus or a major identifiable characteristic of teaching and learning mathematics in the East, in contradistinction to the situation in the West where classroom teaching and learning experiences are strongly geared to inventiveness and creativity.

Commenting on the distinctions between the "East" and "West" as they affect mathematics education in both cultures, Watkins and Biggs (2005) used the paradox construct to address the issue, stating that reflective repetition and effort are strong culture-influenced attributes of Eastern education as opposed to ability, which is the overriding characteristic of Western education. The implied paradoxical question seems to be: Which paradigm (Eastern or Western) yields the better educational outcome? This question remains the focal point of discourse in comparative studies of the East-West educational paradigms. Leung (2001) made a notable contribution to this discourse when he described key contrasts between East Asian and Western traditions in mathematics education using six oft-cited dichotomies. These dichotomies, cited by and adapted from, Peng, Ezeife, and Yu (2018), are summarized in Table 1.

Table 1 Leung's six dichotomies/contrasts between Eastern and Western traditions in mathematics education

S/N	*Type/name of dichotomy*	*Practice in Eastern education*	*Practice in Western education*
1	Product (content) versus Process	Basic knowledge/skills put in the foreground	Focus on the process of doing mathematics
2	Rote learning versus meaningful learning	Rote learning/memorization seen as legitimate and necessary in learning	Emphasis on the necessity to understand the phenomenon before internalization
3	Studying hard versus pleasurable learning	Studying viewed as a serious endeavor that relies on hard work and perseverance	Child put in the middle of learning such that the child enjoys a meaningful learning process
4	Extrinsic versus intrinsic motivation	Extrinsic motivation emphasized as complementary to intrinsic motivation	Intrinsic motivation valued more than extrinsic motivation in mathematics learning
5	Understanding of the nature of the teacher	Whole-class teaching with the teacher as the role model is seen as very important	Stronger focus on individualized learning that lays emphasis on independence/individualism
6	Understanding of the role of the teacher	Teacher as a scholar with deep subject-matter or content knowledge	Teacher as a facilitator with sound pedagogical competencies

In totality, a large number of research studies on the East-West educational paradigms have made significant comparisons of the approaches currently adopted in the teaching and learning of mathematics in both educational systems. These studies have been limited, to a large extent, to discussing the commonalities and differences of mathematics education in both the East and West at a broad or general level, without delving into specific areas of mathematics education. In this study, the researchers deviated from this general trend and focused instead on a specific concept in mathematics education—problem solving.

The overall purpose of the study was not to seek an answer to the lingering question about which educational paradigm (the East or West) could be regarded as yielding the better educational outcome in mathematics education. Rather, the study focused on problem solving, which is generally regarded as a key concept in mathematics teaching and learning. This is because problem solving is a mathematical activity distinct from other mathematics teaching/learning areas such as mathematical concepts, algorithms, and theorems, and it has long been a key component of school mathematics.

One of the fundamental goals of school mathematics has been the development of students' abilities to solve problems. The teaching of problem solving has always been the focus of both research and practice. With this in mind, the following research question was formulated to guide the study: What are the commonalities and differences between two Canadian and Chinese elementary schools in the teaching of problem solving? This research question was addressed by analyzing the data collected from several research interactions and engagements between the two schools.

2 Research Context

2.1 The Reciprocal Learning Between Canada and China Project

In a general context, the concept of reciprocity suggests or indicates a situation which embodies and promotes "a mutual exchange of advantages or privileges" (Barber, 1998, p. 1204). The reciprocal learning approach adopted in the current Canada–China project utilizes an interactive dialogue format which involves educational institutions, researchers, school administrators, teachers and students in Canada (the Western paradigm), and China (the Eastern paradigm). Designated as the *Reciprocal Learning Partnership Project in teacher education and school education between Canada and China*, the seven-year longitudinal project is funded through a Partnership Grant of the Social Sciences and Humanities Research Council of Canada [SSHRC] (Xu & Connelly, 2016). Two Canadian and five Chinese universities, two Canadian school boards, and over forty Canadian and Chinese schools constitute the participants in this international study.

The study has the overall aim of building a strong knowledge base geared to the understanding and implied comparison of the educational views, classroom practices, and activities of practitioners in the Canadian and Chinese educational structures and systems in a reciprocal, engage-and-learn-from-each-other fashion. The reciprocal learning relationship is actualized through the involvement of a well-oiled composite team of researchers from the participating Canadian/Chinese universities, elementary and secondary school teachers in Canada and China, school administrators and district advisors also from both countries. The project is structured in a four research-team format comprising the general education team, mathematics education team, teacher education team, and science education team. The primary goal of the mathematics team is to provide a platform for Canadian mathematics teachers to build long-lasting and meaningful relationships with their Chinese counterparts, thereby enabling teachers from both countries to learn about one another's cultural perspectives on mathematics education and how these perspectives affect not only their teaching, but also students' learning (Zhu & McDougall, 2017).

The mathematics team further created two mutually interactive research sub-teams and Sister Schools with each pair of sub-teams and Sister Schools working on defined topics, concepts, curriculum, and implementation practices in mathematics education in the participating countries—Canada and China. This study reports the results from one of the research teams arising from interactions with school principals, other school administrators, and mathematics teachers from a pair of Grade 2 classes in two Sister Schools.

2.2 The Canadian and Chinese Elementary Schools' Context

To ensure the anonymity guaranteed to the participating schools, pseudonyms were used in the study. Thus, *School A* and *School B* were used to stand for the Canadian school and Chinese school, respectively. Located in Windsor, Ontario, Canada, School A is a public school (Junior Kindergarten to Grade 8) with a student population of 400+. The school has four primary classrooms, two early years' classrooms, and teachers for special education and learning support. The school has made character education its cornerstone. To celebrate character traits, such as respect, caring, self-discipline, the school holds monthly assemblies. The school supports mathematics by posting school-wide problems of the week, posting problem-solving strategies in classrooms, and using a school-wide problem-solving model. The school has an extensive interschool and after school program in sports, robotics, coding, games, and reading. Technology is present in classrooms, particularly Smart Boards, iPads, laptop computers, and data projectors.

School B is located in Chongqing, China. It is a public school with 3000 students ranging from Junior Kindergarten to Grade 6. This is the most prestigious primary school in Chongqing. There are 72 classes and 40 mathematics teachers in the school. The mathematics teachers normally would teach two lessons per day in two separate classes. There is an extensive interschool program, focusing on many of the same activities as in School A. The main technology devices in the school are whiteboards and data projectors. There are some residential students, and so there are evening classes and teacher-helpers for these students.

3 Data Collection and Data Analysis

3.1 Data Collection

A series of data collection techniques were effectively tapped and utilized to source out research data from the two participating schools. These included Skype meetings in which teachers, researchers, school administrators, and students actively participated and shared ideas and actual classroom teaching/learning experiences. Furthermore, online and electronic data-collecting interactions such as e-mail transmissions, *Pepper* and *Blackboard* software systems, *WeChat*, and *Tencent QQ* were extensively employed as research tools. The employment of these multiple techniques for data collection enabled and facilitated the sharing of teaching materials and resources, the creation of teaching videos, and joint observation of real-life classroom sessions and activities.

Such life lessons were then subjected to "debriefing and discussion" sessions during which participating teachers, researchers, school administrators, and students fully shared their observation notes and made remarks about the strengths of observed lessons and/or areas that could be improved upon, especially in lessons that focused on problem solving, and related teacher approaches/student learning skills.

Out of a total of 10 Skype meetings held between the Sister Schools over the two-year duration of this study, six (60%) of the meetings were devoted to problem-solving engagements and activities. During some of the meetings, teachers would teach demonstration lessons on problem solving using specially selected teaching aids, manipulatives, and at times, online teaching resources. For instance, one of such lessons was so designed that it involved the students in the three-phase teaching/learning technique or cycle comprising exploration, concept introduction, and concept application (consolidation) phases (Karplus et al., 1977). Essentially, at the exploration phase, the students were engaged in exploratory hands-on activities and learning experiences aimed at, and leading to, the introduction of the theme (the concept of 'sharing' or division) on which the lesson focused. During the second phase (concept introduction), the students were given exercises that entailed the utilization of the ideas they had gathered from (and the experiences they had been exposed to during) the exploratory activities which they had earlier engaged in. Working in groups, the students were asked to share or divide whole cakes, pizzas, candies, rectangular/square tiles, and lengths of ropes/strings into smaller equal parts. At the concept application or consolidation phase of the lesson, the students were challenged to suggest other everyday items they could divide, first into equal segments, and then into unequal pieces. Thus, the students were, at this final (consolidation or application) phase, exposed to applying or transferring the skills and experiences they had acquired during the first two phases of the teaching-learning interaction in actual problem-posing and problem-solving situations in real-life settings.

As part of data collection, teachers and researchers discussed several issues in mathematics education, with particular reference to problem solving. Some of the issues discussed include use of textbooks; use of aids, resources, and manipulatives; and preparation and use of lesson notes, teaching units, and lesson plans. In addition, the researchers engaged the teachers from both Canada and China in formal discussions with regard to their classroom experiences and student involvement in the teaching of problem-solving lessons. These discussions were aimed at sourcing out actual teacher practices in the delivery of problem-solving lessons with the attendant student responses.

Altogether, four such taped discussions were held, each of which ranged in duration from one to one-and-half hours. The research graduate assistants also took detailed notes during all of the discussions and elaborated on them after each discussion session. Later, the discussion notes were entered into a Word document. This enabled the researchers to dwell on and duly review the notes and delineate valuable research data from them.

Furthermore, the researchers made six data-gathering research field trips to the Project Sister School in China and five trips to the Canadian Sister School. During the trips, teaching/learning facilities in the schools were examined and useful pieces of research data were tapped from them. Facilities examined included school libraries, mathematics textbooks, teaching materials and teachers' notes, classroom set-ups, sample "problems-of-the-week" on classroom walls and school hallways, school manipulatives' stores, and teacher-made improvised materials/devices used for problem-solving lessons.

3.2 Data Analysis

In the analysis of the data, the researchers adopted a data coding technique modeled on grounded theory (Corbin & Strauss, 2008). The adoption of this approach was informed by the fact that this specific approach aligned closely with the main goal or thrust of the study which is the detailed explication of "an educational process of events, activities, actions, and interactions that occur over time" (Creswell, 2014, p. 432). This detailed explication was specifically aimed at understanding the commonalities and contrasts in the teaching of problem solving in the two research Sister Schools.

Adopting a stepwise method in the data analysis process, the researchers first systematically reduced the data by summarizing the relevant information on mathematical problems and problem solving that had been compiled from the multiple data collection techniques utilized in the data collection segment of the study. Secondly, those parts of the interview and Skype meeting notes that were in Chinese were translated into English. Coding categories were then developed and consequently used to code the data and sort them by category. Data within each category were compared to one another, and findings from different data sets were also compared. The researchers then proceeded to analyze the data individually, but jointly discussed the findings as an interactive team, and finally developed a mutually acceptable interpretation of the data as recorded in the research results and findings.

4 Findings

4.1 Commonalities in the Teaching of Problem Solving Between the Two Canadian and Chinese Schools

Two recurring commonalities regarding the teaching of problem solving in the two Canadian and Chinese Sister Schools emerged from our analysis. The first is the deliberate choice and use of problems that have connections with real-life situations, and the second is the obvious inclination of the teachers to encourage students to use multiple strategies to solve problems. These themes were abundantly discernible from the classroom interactions that occurred in all teaching and learning engagements on problem solving in the two schools all through the duration of the research study.

Our results show that all of the Canadian and Chinese teachers in the two research schools recognize the immense value of teaching mathematical problem solving in a real-life context. The use of this approach falls in line with the increasing emphasis being placed on the application of mathematics in real-life situations in curriculums in most countries of the world. Yeo (2007) drew attention to the fact that real-life tasks are often readily familiar, meaningful, and relevant to students; hence, they can serve as solid stepping stones and strong anchor bases for thinking about important mathematical concepts.

The results of the international study conducted by Bryan, Wang, Perry, Wong, and Cai (2007) firmly supported the stance that there is immense value in teaching problem solving in a real-life context. In the study, the authors interviewed mathematics teachers from four regions of the world, namely Australia, the United States of America (USA), Mainland China, and Hong Kong. As recorded in the following responses, the interviewed teachers hold the strong view that mathematics—particularly its problem-solving aspect, component, or dimension—has a string of deeply embedded, inseparable, and intertwining relationships with societal issues and affairs in the world in which both the teacher and the taught, live and function. The existence of these deep-seated relationships strongly suggests that mathematics (with particular emphasis on problem solving) should be taught in a real-life context, for maximal benefits. A sample of the teachers' responses, quoted from (Bryan et al., 2007), and listed below, succinctly affirms their stance:

> For Australian teachers, mathematics is one of those essential subjects that allows us to function in the world. For Mainland Chinese teachers, mathematics is practical in daily life and can help people solve real life problems in an efficient way. It is a science as well as a necessary tool for life. For teachers from Hong Kong, the practical significance of mathematics constituted a salient theme in the teachers' response. In daily life, [a] child may face problems in books. When they grow old, they use it in buying [a] house. I think that we learned some skills and method of calculation, then apply them in life to solve problems continually. For U.S. teachers, mathematics could provide a new perspective for looking at the world: "I see it as a tool in order to solve problems... But it's a tool that enables people to

> do things or to reach goals that they have. The substance of mathematics would be things like a set of rules, a set of methods that allow me to achieve goals or achieve things I'm trying to do or other people are trying to do." (p. 330)

4.2 Multiple Methods of Solving a Problem

The expressive dictum or maxim that states that "variety is the spice of life" has direct and weighty implications for the teaching and learning of problem solving in mathematics education. It is the variety the students are exposed to with regard to the availability and choice of alternative methods of solving a given mathematical problem that spices up the students' interest in the problem-solving exercise or engagement. As Schoenfeld (1985) emphasized, the degree of student interest, and active participation or engagement, is usually heightened when they are conscious or aware that a problem under consideration may be solved in multiple ways. This is in agreement with the widely acknowledged assertion that the most important aspect or part of teaching is not to stuff learners with facts, but to awaken in them the curiosity to learn.

Tisdell (2018) drew attention to the felt need to expose learners to multiple and diverse approaches that may be available for solving given problems because this would enable individual learners to choose and adopt an approach that is best suited to, and rhymes with, their learning styles and preferences. Stressing the advantages of the use of multiple methods in mathematics problem solving, Tisdell (2018) cited several researchers on problem solving who all share the informed view that:

> an understanding of multiple solution methods is essential for mathematical comprehension due to this approach enabling new connections to be made between seemingly separate areas of mathematics, forming new knowledge networks; or for new material to be embedded within schemas familiar to the learner. (p. 793)

Similarly, Leikin and Levav-Waynberg (2008) made the notable assertion that: "solving problems in multiple ways contributes to the development of students' creativity and critical thinking" (p. 234). Our results indicate that teachers in both research Sister Schools are aware of the importance of using different strategies in mathematics problem solving. We drew this conclusion based on our data, which glaringly show that the teachers who participated in the study regularly encouraged students to use multiple strategies to solve problems in all class sessions, interactions, and engagements when problem solving was taught.

4.3 Contrasts in the Teaching of Problem Solving Between the Two Canadian and Chinese Schools

Three easily identifiable contrasts regarding the teaching and learning of problem solving in the two Canadian and Chinese schools emerged from the analysis of this study. These are teachers' teaching strategies, students' learning tendencies, and sources of mathematical problems.

4.3.1 Teachers' Teaching Strategies and Their Effects on Student Learning

Teachers' teaching strategies varied remarkably when the strategies adopted by Chinese teachers were compared to the strategies used by their Canadian counterparts. Whereas Chinese teachers emphasized problem-posing and utilized it as a critical component in the delivery of their problem-solving lessons, Canadian teachers employed a systematic approach (which usually involves a step-by-step procedure) to mathematical problem solving. Several research studies (Cai & Hwang, 2002; Guvercin & Verbovskiy, 2014; Lavy & Shriki, 2007) have all described the advantages that usually accrue to mathematics learners who acquire problem-posing skills in mathematics education. These skills often solidify and sharpen students' problem-solving abilities in general and confer direct ownership and a valued sense of accomplishment to the students when they successfully create viable problems on their own.

In the spirit of reciprocity of the project, both the Chinese and Canadian teaching teams in this study encouraged participating students to pose mathematics problems, which the students then challenged one another to solve, sometimes in group settings. One of such student-posed problems, based on *Rectangular models of fractions/ordering of fractions* (and depicted in Figs. 1), reads: *Two students, Laura and James, were shown two equal rectangular chocolate bars each of which is divided up into fractional parts. The first bar* (Fig. 1a) *is divided into nine equal parts, while the second bar* (Fig. 1b) *is divided into eleven parts.*

Question: If 7 out of the 9 fractional parts are cut out from the first bar (Fig. 1a), and 9 of the 11 fractional parts are cut out from the second bar (Fig. 1b), which cutout (the one from Figs. 1a or 1b) contains more

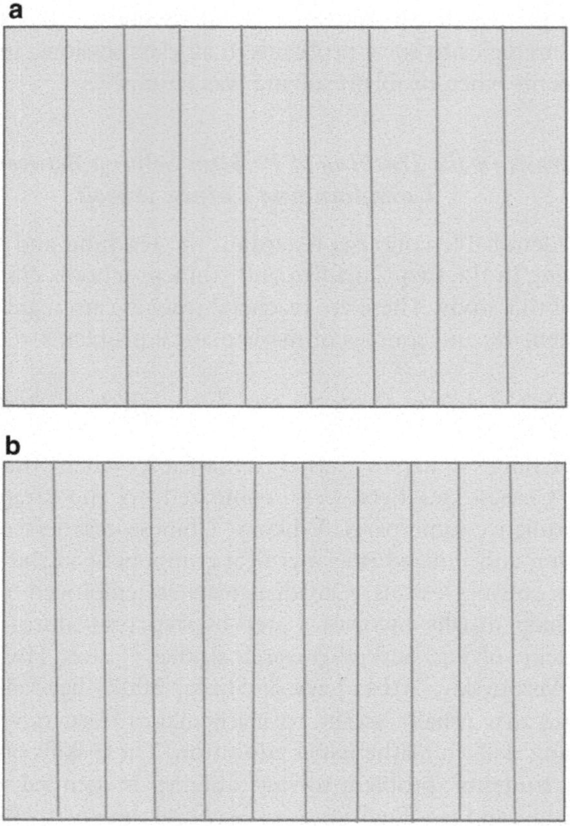

Fig. 1 **a**. A rectangular chocolate bar divided into 9 equal fractional parts; **b**. A rectangular chocolate bar divided into 11 equal fractional parts

quantity of chocolate? If Laura chose the cutout from Fig. 1a while James chose the cutout from Fig. 1b, which student was right? Why?

4.3.2 Class Involvement and Interactions Emanating from Student-Posed Problems

As researchers observing classroom activities and learning sessions during the study, it was interesting and informative to record the almost palpable zeal exhibited by the participating students as they enthusiastically set to work solving each of the problems posed by their classmates. Later, in the same ebullient manner, they jointly discussed and shared the approaches they adopted in solving each of the posed problems. The interest generated by the problem-posing sessions was remarkable. The students' eyes seemed to have been suddenly opened with regard to the very possibility that they could pose and solve "their own" mathematics problems.

It was also noteworthy and instructive to observe that most of the problems posed by the students were drawn directly from familiar activities, events, and experiences they usually encounter in their day-to-day lives. This finding from our study firmly supports the stance of Lavy and Shriki (2007) who asserted that the coveted spirit of ownership acquired by students when they pose viable problems themselves "results in a high level of engagement and curiosity, as well as enthusiasm towards the process of learning mathematics" (p. 130).

Students' Different Learning Tendencies

One of the six dichotomies or contrasts highlighted by Leung (2001) with regard to Eastern and Western traditions in mathematics education dwelt on the topic of whole-class teaching versus individualized learning practices. As detailed in Table 1, whole-class teaching with the teacher as the role model is seen as very important and practiced in the East, as opposed to the situation in the West where there is a stronger focus on individualized learning that lays emphasis on independence/individualism. However, the findings of our study revealed a situation contrary to this regularly expressed and commonly accepted status quo in the East and West with regard to mathematics education, specifically the theme of our study—mathematics problem solving. This is because, in this study, it was observed that, when given a mathematics problem during class teaching/learning engagements, the Chinese students, contrary to expectation, tended to solve it individually, whereas the Canadian students would bunch together to solve the given problem cooperatively. These observations were comprehensively recorded in the teaching notes one of the Chinese teachers shared with the researchers (Peng, Ezeife, & Yu, 2018) in which the teacher stated:

> Cooperative learning seems a natural learning habit to the Canadian students. When given a problem, they immediately gather together. This action seems very ordinary and simple, but reflects the idea that cooperative learning is deeply rooted in the students' minds. (p. 12)

Sources of Mathematical Problems
The selection and use of problems in problem-solving teaching/learning sessions in the two research Sister Schools were found to reflect the deep-rooted beliefs held by the teachers who delivered the lessons. This, to a large extent, was deciphered to be a product of the belief systems, educational paradigms, practices, and goals in the East and West. As pointed out by Watkins and Biggs (2005), teachers in the East lay a lot of emphasis on effort and preparing their students for competitive examinations that would lead them to success in the future. In the field of mathematics education, most of the textbooks in use are structured with this success-geared orientation in focus, and the Chinese mathematics teachers in our study were seen to routinely select problems from different books for their students.

On the contrary, the Canadian teachers in the study largely utilized technological devices such as iPads, computer-based software, spinners, counting devices, and other hands-on manipulative contraptions to source out problems for their students. Thus, our finding in the study indicated that the participating teachers differed in and from where they sourced the mathematics problems they used in their problem-solving classes. This is a difference which, from our analysis, we believe is largely attributable to the teachers' differing orientations and practices in the broader field of mathematics education in the East and West, not just in mathematics problem solving, the focus or theme of our study.

5 Conclusion

Through this interactive, participatory, cooperative, and reciprocal learning-oriented study, the researchers have come up with new knowledge on the commonalities and contrasts in the teaching of problem solving in two research elementary schools in two countries which differ drastically not only culturally, but also geographically—China (in the 'East') and Canada (in the 'West'). Based on our findings, we came to the conclusion that there is a degree of commonality across the two schools in the teaching of problem solving, with a common emphasis laid on having connections with real-life situations, and encouraging students to use multiple strategies to solve problems.

On the other hand, there are some differences between the two schools with regard to teachers' teaching strategies, students' learning tendencies, and sources of mathematical problems. The summary of our findings indicates that:

- In the two schools, students were encouraged to use multiple strategies or methods when solving problems;
- Approaches used by the students were impacted on, or influenced by, the methods they are taught and hence usually (commonly) adopt.
- Canadian and Chinese teachers differ slightly in their perspectives on, and approaches to, teaching problem solving. For example, problem-posing is emphasized by Chinese teachers whereas Canadian teachers prefer a systematic approach when teaching problem solving.
- Canadian teachers mentioned that they would invite students to discuss possible methods of solving a given problem, but they would not suggest the optimal methods; instead, the students decide the methods they like. On the contrary, Chinese teachers mentioned that they would discuss the advantages and disadvantages of the different potential methods and then give suggestions on when to use the different methods.

6 Further Studies

The results and findings of this study raise some pertinent questions for future research such as why do the discerned differences exist and how can they be gainfully addressed and/or possibly resolved? For instance, why is problem-posing emphasized by Chinese teachers whereas Canadian teachers prefer a systematic approach when teaching problem solving?

Even though this study was conducted for a seemingly long period of two years, the researchers consider that duration as just enough time to scratch the weighty issue merely on the surface. It is our belief that subsequent future studies should more deeply explore these differences over a longer time duration through continuing interactions among students, teachers, and researchers in the two countries—Canada and China. It is hereby suggested that such future studies should continue using the concept of Sister Schools and communities, increasing the number of schools, participating primary and secondary school teachers, students, scholarly communities, and other education-positioned stakeholders, to the extent that available funding and associated resources would permit.

It is possible that more data could emerge from these suggested future studies, which would hopefully reveal more thematic features that typify the teaching and learning of mathematics in Canada and China. This is the suggested direction future studies could pursue—a direction which would enable researchers to explore further in the fertile field of mathematics education—a field in which the enthralling mixture of curriculum, pedagogical issues, and practices consistently engage and challenge the active researcher's mind in the knowledge-searching process of doing and undergoing.

Acknowledgement The researchers gratefully acknowledge the grant from the Social Sciences and Humanities Research Council (SSHRC) of Canada, which funds the larger Reciprocal Learning Canada-China Partnership Project (898-2021-1011, Xu & Connelly, 2019), and on which this mathematics-focused study is based. The National Social Science Foundation in China (17XMZ032), which also partially funded the study, is also hereby appreciated.

References

Barber, K. (Ed.) (1998). *The Canadian Oxford Dictionary*. Toronto: Oxford University Press.

Bryan, C. A., Wang, T., Perry, B., Wong, N., & Cai, J. (2007). *Comparison and contrast: Similarities and differences of teachers' views of effective mathematics teaching and learning from four regions*. Retrieved from https://www.researchgate.net/publication/227264205_Comparison_and_contrast_Similarities_and_differences_of_teachers%27_views_of_effective_mathematics_teaching_and_learning_from_four_regions.

Cai, J., & Hwang, S. (2002). Generalized and generative thinking in U.S. and Chinese students' mathematical problem solving and problem posing. *Journal of Mathematical Behavior, 21*(4), 401–421.

Corbin, J. M., & Strauss, A. (2008). *Basics of qualitative research: Techniques and procedures for developing grounded theory* (3rd ed.). Los Angeles, CA: Sage.

Creswell, J. W. (2014). *Educational research: Planning, conducting, and evaluating quantitative and qualitative research* (5th ed.). Upper Saddle River, NJ: Pearson Education.

Guvercin, S., & Verbovskiy, V. (2014). The effect of problem posing tasks used in mathematics instruction to mathematics academic achievement and attitudes toward mathematics. *International Online Journal of Primary Education, 3*(2), 59–65.

Kaiser, G., & Blömeke, S. (2013). Learning from the Eastern and the Western debate: the case of mathematics teacher education. *ZDM-The International Journal on Mathematics Education, 45*(1), 7–19.

Karplus, R., et al. (1977). *Science teaching and the development of reasoning (physics)*. Berkeley, CA: University of California Press. Retrieved from https://onlinelibrary.wiley.com/doi/pdf/10.1002/tea.3660140212.

Lavy, I., & Shriki, A. (2007). Problem posing as a means for developing mathematical knowledge of prospective teachers. In J. H. Woo, H. C. Lew, K. S. Park, & D. Y. Seo (Eds.), *Proceedings of the 31st conference of the international group for the Psychology of Mathematics Education*, Vol. 3 (pp. 129–136). Seoul: PME. Retrieved from http://www.emis.de/proceedings/PME31/3/129.pdf.

Leikin, R., & Levav-Waynberg, A. (2008). Solution spaces of multiple-solution connecting tasks as a mirror of the development of Mathematics teachers' knowledge. *Canadian Journal of Science, Mathematics and Technology Education, 8*(3), 233–251.

Leung, F. K. S. (2001). In search of an East Asian identity in mathematics education. *Educational Studies in Mathematics, 47*(1), 35–51.

Peng, A., Ezeife, A. N., & Yu, B. (2018). Reciprocal learning in mathematics education: An interactive study between two Canadian and Chinese elementary schools. *Comparative and International Education, 47*(1), 1–14.

Schoenfeld, A. H. (1985). *Mathematical problem solving*. Orlando, FL: Academic Press.

Tisdell, C. C. (2018). Pedagogical alternatives for triple integrals: moving towards more inclusive and personalized learning. *International Journal of Mathematical Education in Science and Technology, 49*(5), 792–801. Retrieved from https://www.tandfonline.com/doi/pdf/10.1080/0020739X.2017.1408150?needAccess=true.

Watkins, D. A., & Biggs, J. B. (Eds.) (2005). *The Chinese learner: Cultural, psychological and contextual influences*. Melbourne: Comparative Education Research Centre (CERC) & Australian Council of Educational Research (ACER).

Xu, S. J., & Connelly, F. M. (2013–2020). Reciprocal learning in teacher education and school education between Canada and China. Unpublished research project proposal. An eight-year standard research grant funded by Social Sciences and Humanities Research Council (SSHRC) of Canada [Grant No. 895-2012-1011].

Xu, S., & Connelly, M. (2016). *Partnership grant reciprocal learning in teacher education and school education between Canada and China*. Retrieved from http://www.reciprocal-learning.ca/pages.php.

Yeo, J. B. (2007). *Mathematical tasks: Clarification, classification and choice of suitable tasks for different types of learning and assessment*. Report No. ME 2007-02. Singapore: Nanyang Technological University.

Zappia, S. (2018). *Major difference between western & eastern thought on education*. Retrieved from https://education.seattlepi.com/major-difference-between-western-eastern-thought-education-2355.html.

Zhu, C., & McDougall, D. (2017, April 27–May 1). Building successful reciprocal learning relationships between Canadian and Chinese mathematics teachers: A researcher's perspective. *Paper presented at the Annual Meeting of the American Educational Research Association (AERA)*, San Antonio, Texas.

CHAPTER 11

Research on the Teaching of Understanding Mathematics Problems in Elementary Schools

Yufeng Luo and Bo Yu

1 Background

Since the beginning of 1980s, the National Council of Teachers of Mathematics has proposed that "problem solving" should be the core of mathematics education (English & Gainsburg, 2016). "Problem solving" has become an important part of mathematics education research and part of the main content of mathematics education in primary and secondary schools in China. Problem solving is seen as an important way of mathematics teaching. In China, cultivating "problem solving" competence has been regarded as one of the goals of compulsory education mathematics curriculum standard (Ministry of Education, 2012).

Y. Luo (✉)
Xixin Primary School, Chengdu, China
e-mail: 2294618534@qq.com

B. Yu
College of Teacher Education, Southwest University, Chongqing, China
e-mail: 1354701002@qq.com

S. C. Zhu et al. (eds.), *Reciprocal Learning for Cross-Cultural Mathematics Education*, Intercultural Reciprocal Learning in Chinese and Western Education,
https://doi.org/10.1007/978-3-030-56838-2_11

However, in the teaching practice of mathematical problem solving, we found that students often fail to understand the problem. This can lead to the problem not being solved, and the student not being able to experience the full exercise of thinking, the consolidation of knowledge, and the improvement of ability from mathematical problem solving (Dai, 1997). It makes us wonder: How do students understand mathematics problems? How should teachers help students understand mathematics problems?

This study was part of a larger reciprocal learning project (Xu & Connelly, 2013) and investigated the teaching of mathematics problem understanding by two elementary school mathematics teachers, one in China and one in Canada. Through the analysis of mathematical problems, representation of mathematical problems, and components of mathematical problems through the reciprocal learning programs in China and Canada, we discuss the strategies and methods for the teaching of understanding of mathematical problems. The cognitive process of elementary school students' understanding of mathematical problems is constructed from the perspective of information processing theory (Dai, 1997; Zhong, 2007).

In this chapter, we describe the teaching elements that affect students' understanding of mathematical problems and the characteristics and significance of their understanding of mathematical problems. In order to advance the goal of solving mathematics problems in compulsory education, we compare the teaching process of solving mathematics problems of two primary school teachers in China and Canada and explore the teaching suggestions that can promote the understanding of mathematics problems by primary school students.

Hopefully, we will provide mathematics teachers with some support in how they can improve students' ability to understand mathematics problems. By learning about student's understanding of mathematical problems and enriching the teaching practice for problem solving, we can better cultivate and improve students' ability to solve problems and improve students' thinking process.

2 Literature Review

This literature review is mainly to provide theoretical support to this study, which embodies in two aspects. The first aspect is to collect the relevant literature on mathematical problems, mathematical problem solving, teaching process, and mathematical problem understanding, in order to understand the current research status of mathematical problem solving and understanding. The second aspect is to collect, analyze, and sort out the relevant teaching theories of mathematics in order to better understand teaching to provide support for the teaching methods of mathematics problem understanding in primary schools.

Teaching is a bilateral activity between teachers' teaching and students' learning. The basic process of students' understanding of mathematical problems is the basis of teaching (Li, 2005). Based on the cognitive process of students' understanding of problems, what kind of teaching support should teachers provide to help students understand problems?

According to the bilateral characteristics of teaching, the analysis of elementary school students' understanding of mathematical problems is mainly approached in two ways. One way is from the student perspective. Teachers must have the attention of the students. The analysis of students' mathematical understanding mainly contains analyses of the thinking characteristics, language characteristics, as well as existing knowledge and experience of primary school students (Zhong & Liu, 2014).

The second aspect is based on the requirements of students' development and the teaching objectives that teachers provide to support students' understanding of problems. Teaching support mainly includes providing additional examples, mobilizing cognitive background, promoting the connections, and guiding students to control the cognitive process, as shown in Table 1. In order to show that the objective of teaching is to function as a guide and offer teacher support, the teaching objectives and teaching support are analyzed as a whole in the teaching analysis of mathematical problem understanding lessons.

The analysis of the teaching of mathematical problem understanding is inseparable from the analysis of the learner. When analyzing the learner, we must examine the thinking development of a certain age, their language development, and the existing knowledge and experience of the learner. Secondly, we then analyze the teaching, which is guided by students' learning goals set out by the problem-solving objectives in the new curriculum standards (Ministry of Education, 2012). Finally, based on the different levels of understanding, we analyze what kind of support teaching needs to provide for students' understanding of problems, mainly including the construction of problem field, cognitive mobilization and construction, generation of connection, and representation of problems.

Table 1 Analytical framework of the teaching of primary school mathematics problem understanding (Liu, 2007; Ma, 2001; Shi, 2015; Zhang & Fan, 2016; Zhu, 2009)

Learning subject	*Thinking characteristics: stage 1 (grade 1–3): concrete thinking dominates; stage 2 (grades 4–6): thinking mode is in transition*				
	Language characteristics: The first stage is from grade one to grade three. The goal of students' language development is to standardize language. The second stage is from grade four to grade six. The goal of students' language development is thinking language				
	Existing knowledge and experience: Students' existing knowledge and experience include mathematical concepts, theories, methods, strategies, etc.				
The teaching goal	*Understanding* / *level*	*Instrumental understanding*		*Relational understanding*	
		Intuitive understanding	*Procedural understanding*	*Abstract to understand*	*Formal understanding*
Teaching support	Problem field	The presentation of the problem is mainly based on images and situations	The problem presents prominent types, with known quantity, unknown quantity, condition, and clear target	The problem is contained in the context, including known quantity, unknown quantity, condition, and target, which needs to be extracted with purpose	Detached from the situation, the problem mainly relies on images or symbols to present, known quantity, unknown quantity, and other elements of the problem are sufficient
	The cognitive background	Experience is the main basis of his understanding	The key to its understanding is to solve equations and algorithms	Existing knowledge and experience as well as language foundation	Mathematical axioms, theorems, concepts, and so on are its main basis

(continued)

Table 1 (continued)

Learning subject	*Thinking characteristics: stage 1 (grade 1–3): concrete thinking dominates; stage 2 (grades 4–6): thinking mode is in transition*				
	Language characteristics: The first stage is from grade one to grade three. The goal of students' language development is to standardize language. The second stage is from grade four to grade six. The goal of students' language development is thinking language				
	Existing knowledge and experience: Students' existing knowledge and experience include mathematical concepts, theories, methods, strategies, etc.				
The teaching goal	*Understanding level*	*Instrumental understanding*		*Relational understanding*	
		Intuitive understanding	*Procedural understanding*	*Abstract to understand*	*Formal understanding*
	Create connection	Mainly the connection between situation and experience	Connections arise between knowledge and experience of the same type of problem	Connections occur between situations and experiences, between students' bases and contents, between known quantities and unknown quantities, and between conditions and objectives	The connection is mainly between the elements of the problem, and between the elements and the existing knowledge and experience
	Problem representation	Generally use image or language representation, visual presentation	Generally, the transformation from image representation to written symbol representation can be realized	To realize the internal or interconversion of written symbolic representation and oral linguistic representation	Transformation from symbolic representation to symbolic representation

(continued)

Table 1 (continued)

Learning subject	*Thinking characteristics: stage 1 (grade 1–3): concrete thinking dominates; stage 2 (grades 4–6): thinking mode is in transition*				
	Language characteristics: The first stage is from grade one to grade three. The goal of students' language development is to standardize language. The second stage is from grade four to grade six. The goal of students' language development is thinking language				
	Existing knowledge and experience: Students' existing knowledge and experience include mathematical concepts, theories, methods, strategies, etc.				
The teaching goal	*Understanding level*	*Instrumental understanding*		*Relational understanding*	
		Intuitive understanding	*Procedural understanding*	*Abstract to understand*	*Formal understanding*
	The control process	The control process of problem understanding mainly depends on the creation of the situation	Control the direction of problem analysis and supervise the process of problem understanding	To mobilize metacognitive participation, modify and improve the understanding until the problem is fully understood	To mobilize the participation of mathematical metacognition and intuitive thinking

3 METHOD

The goal of this study is to find out how teachers from both countries teach problem understanding in class and to analyze the differences and existing problems in the teaching of mathematical problem understanding. We observe the mathematics problem understanding lessons taught by teachers in elementary schools in China and Canada. We use the horizontal comparison method to compare the teaching of mathematics problem understanding in primary schools in China and Canada from both macroscopic and microscopic aspects, so as to find out the similarities or differences in the teaching of cognitive processes for the understanding of mathematics problems.

4 Case Study of Mathematics Problem Understanding Teaching for a Primary School Teacher in Chongqing, China

In the classroom teaching of the topic "solving the volume of irregular objects", teachers and students generated a lot of problems and answers through the question-and-answer communication. This study extracted three representative mathematical problems and analyzed their teaching support. The analysis chart is shown in Table 2.

There are different levels of understanding mathematics and low-level understanding can also be meaningful. Students can consolidate new knowledge through procedural understanding and learn how to think and solve such problems. Formal understanding is the combination of procedural understanding and abstract understanding. Most problems that need formal understanding need to be based on mathematical concept knowledge and problem-solving strategies, especially students' logical thinking.

In this class, teacher G, on the basis of students' prior knowledge, uses mathematical problems of different levels to support students' understanding of mathematical problems. The lesson is very focused on the teaching goal. The teacher teaches in a planned and purposeful manner with mathematics problems that are provided through the whole teaching lesson. In this case, teacher G's teaching of understanding mathematical problems fully mobilizes students' prior knowledge and experience to form necessary connections between learning subjects and problems, as well as between various elements of problems. He guides his students to use words, images, and symbols for representation. In this class, the students play a proactive role in finding the volume of irregular objects and analyzing such problems.

5 Case Study of Mathematics Problem Understanding for Primary School Teachers in Windsor, Canada

"How many post-it notes can fill file cabinets?" This question is posed by teacher C of the primary school in Windsor, Canada. It mainly focuses on how to calculate the bottom surface area of cuboid wardrobe. The

Table 2 Teaching support provided by G teachers in Chongqing R primary school based on teaching objectives

Teaching objectives / *Teaching support*	*Program understanding level*	*Abstract understanding level*	*Formal understanding level*
Mathematical problems	Problem 2: Dad put a fake mountain stone in a cuboid bathtub with an area of 51 cubic decimeters at the bottom, and the water rose 3 cm. How big is the volume of the fake mountain stone?	Problem 1: There are two objects in the teacher's box, one is plasticine, and the other is potato. Can you calculate their volume? How do you calculate it? What is the size of the potato? (provide rectangular container, cylindrical measuring cup, ruler, water, and potato)	Problem 4: Find the volume of the large sphere 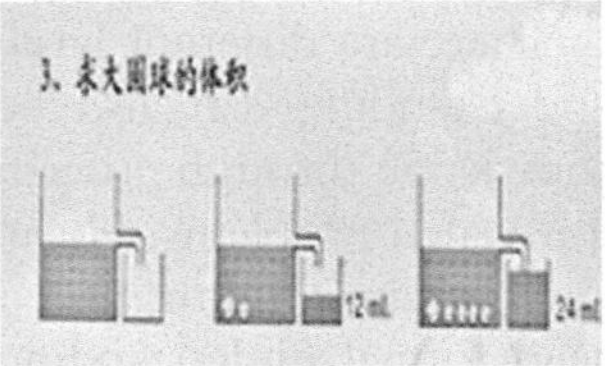
Problem field	The problem is contained in the situations created by the teacher, including known quantities: the bottom area of a cuboid, the height of water rising; unknown quantity: volume of pseudo-rock; conditions: rockery into the bathtub; and objective: to calculate the volume of rockery stone	The problem is contained in the situation created by the teacher, including known quantity: measuring cup, cuboid; unknown quantity: length, width, and height of cuboid, volume of potato; conditions: concrete objects; and objective: to calculate the volume of potato	The problem is out of situations, which is mainly presented by images or symbols. The known quantity: After the large container is put into a large one and a small one, the small cup shows 12 ml liquid; after the large container is put into a large one and a small one, the small cup shows 24 ml liquid; unknown quantity: volume of small ball and big ball; and objective: to find the volume of large round ball

(continued)

Table 2 (continued)

Teaching objectives / Teaching support	*Program understanding level*	*Abstract understanding level*	*Formal understanding level*
The cognitive background	Students have learned the unit conversion, the calculation of the rectangular area, and the volume of a cuboid and have a certain understanding of solving the volume of irregular objects	Students have learned the calculation of the area of rectangle and square, the volume of cube and cuboid, etc. They have mastered the relevant mathematical concepts such as area and volume; also known that things have volumes	Students have mastered the method of finding the volume of an object by drainage and understand the principle of drainage—the volume of an object rising into a container of liquid is the volume of the object, have already had the experience of using transformation thought, and so on
Create connection	Before solving this problem, students have gained experience in using the "drainage method" to find the volume of irregular objects, that is, the connection creates between the same types	The plasticine and potato in the situation are connected with daily life experience. The volume can be calculated by measuring the length, width, and height of a cuboid with a ruler. There is a connection between cuboid volume, measuring cup, and potato volume	Connections are created between various elements, such as the connections between the large sphere and the small sphere, the volume of the sphere, and the volume of the liquid in the small cup, as well as between each element and the existing knowledge and experience

(continued)

Table 2 (continued)

Teaching objectives / *Teaching support*	*Program understanding level*	*Abstract understanding level*	*Formal understanding level*
Problem representation	Students can directly represent with written symbols	To be able to orally express the understood problem, especially with mathematical symbols, it is also able to interpret written symbolic representations using oral language	Students can think directly about images to realize the transformation from symbolic representation to symbolic representation
The control process	Students already have the corresponding understanding ideas, can control the direction of problem analysis, and supervise the problem understanding process	Students can mobilize metacognitive participation to modify and improve the understanding until the problem is fully understood	To mobilize the participation of mathematical metacognition and intuitive thinking

teaching of this class has the characteristics of Canadian teachers' mathematics teaching. The teacher did not directly give the length, width, and height of the wardrobe, but first asked the students to estimate the length, width, and height of the wardrobe through daily life experience and prior knowledge, emphasizing the importance of estimation for solving problems. By posting post-it notes, the teacher indirectly suggests that students can solve problems using that technique and provides students with a learning list for problem solving.

Teacher C provides students with information, and students cooperate and communicate with their deskmates under the guidance of the teacher. Together, the students and teacher explore the surface area of objects and continuously build understanding of the mathematical problems in class, so as to acquire the application of knowledge, improvement of ability, and mastery of thinking methods.

In the "how many post-it notes can fill file cabinets?" activity, teacher C focuses on raising a mathematical problem, analyzing a mathematical problem, and solving a mathematical problem. This study extracts this problem and analyzes the teaching support for understanding mathematical problems provided by it. The analysis chart is shown in Table 3.

This problem may take two hours in an Ontario classroom. It is not possible to take this much time on a single lesson in China. Teacher C's way of putting forward questions left a deep impression on students. Pictures and video display completely attracted students' attention, and students unconsciously acquired the key information. Teacher C emphasized the participation of students in the process and the cooperation and communication between partners and does not pay special attention to the learning of the students.

Teacher C provides students with the support of teaching, mainly reflected in the study sheets provided, and the key information was explained by the teacher. However, the students did not quickly create the connection between the post-it notes and the length, width, height, and surface area of the cabinet, which is the key reason that most students fail to solve the problem correctly. This provides a thinking direction for the teaching research of mathematical problem understanding.

6 Findings and Teaching Suggestions

The main purpose of this research is to provide examples of primary mathematics teachers teaching problem solving, to cultivate students' understanding of mathematics problems, and to improve the effectiveness of mathematics problem solving. We will present some findings and some teaching suggestions for understanding mathematical problems.

This study is based on the teaching of the understanding of mathematical problems in "solving the volume of irregular objects" by teacher G of a primary school in Chongqing, China, and "how many post-it notes can fill file cabinets?" activity by teacher C of a primary school in Windsor,

Table 3 Teaching support provided by Canadian T primary school teachers based on teaching objectives

Teaching objectives / *Teaching support*	*Intuitive comprehension level*	*Abstract understanding level*
Mathematical problems	Problem 1: What is the estimated length, width, and height of the cabinet?	Problem 2: If the length of the cabinet is 90 cm, the width is 45 cm and the height is 180 cm, use sticky notes to stick the length of the cabinet, 12 pieces in a row and 4 rows are stuck, and the last one says 48. What is the length, width, and height of the cabinet? How many post-it notes can cover the entire cabinet?
Problem field	The question is contained in a photo of a file cabinet provided to the students	The problem is contained in the situation and presented in the form of pictures and video. The problem still includes the known quantity: the length, width, and height of the cabinet, 12 stickers in a row of the cabinet length; unknown quantity: the length of a convenient side and the area of a post-it note, and the purpose is to calculate how many post-it notes can fill the file cabinet
The cognitive background	Daily experience with cabinet length, width, and height	How to measure the length of other objects without a standard ruler; students can calculate the area of the figure and the concept of the guiding surface area

(continued)

Table 3 (continued)

Teaching support \ Teaching objectives	*Intuitive comprehension level*	*Abstract understanding level*
Create connection	Create connections between pictures and experiences	The connection between the students' knowledge of mathematical concepts and the elements of the problem, known quantity (length, width, and height of the cabinet, cabinet long a row of 12), and unknown variables (a convenient side length, the area of a post-it note), condition and goal (how many post-it notes can be plastered with filing cabinets)
Problem representation	General use image or language representation, visual presentation	To express orally what is understood and to represent it in mathematical notation; it can also use oral language to interpret written symbol representation and realize the transformation from symbol representation to symbol representation
The control process	The process of regulating problem understanding mainly relies on the provided pictures	Students can mobilize metacognitive participation to modify and improve the understanding until the problem is fully understood

Canada. The teaching comparison of the understanding of math problems between two elementary school math teachers in Canada and China yields five findings.

Our first finding, based on the teaching objectives, shows that both teachers emphasized that students explore mathematical problems through observation, thinking, and hands-on operation and focus on cultivating students' sense of cooperation. It was found that the Chinese teacher G focused on students' methods of understanding problems and corresponding knowledge through a series of hands-on activities. The

Canadian teacher C focused on a series of thinking processes of students facing mathematics problems. To teacher C, whether or not the student solves the problem is not the most important. There is no step-by-step guidance for understanding mathematical problems, and the teacher only reminds the students of key information. The understanding and discovery of solutions to problems require students to try by themselves.

The second finding is related to the creation of mathematics problems. The study found that teachers from both China and Canada have provided a problem situation for students. For primary school students, creating problem situations can bring students closer to learning and real life, realize the practical value of mathematics, and help them to put what they have learned into practice. The problems created by Chinese G teachers are of an advanced level and relevance. The advanced level is reflected in the fact that the proposed mathematical problems can be applied to the different levels of understanding among students. The mathematics problem created by Canadian teacher C is an open problem. It was found that the understanding of the problem does not require the student to be restrained to one kind of solution nor would the "right" answer by provided to the students. They hope students can try and find it themselves.

The third finding is that both teachers focus on scaffolding for students' understanding and activating the prior knowledge and experience. However, there are differences in the methods of transferring prior knowledge and experience. Teacher G in China asks "can you find the volume of a cuboid?", which reviews the prior knowledge and also lays the foundation for the acquisition of new knowledge. Canadian teacher C did not directly give students the knowledge needed to understand or solve this problem, but guided them to understand some general concepts, such as "estimation" and "covering the whole cabinet". It was obvious that there was no review of the prior knowledge, and it was difficult for students to find the junction between prior and new knowledge.

The fourth finding is that there is a creation of the connection between the student and the problem. The effect of the connection is different in the two classrooms because of how the teaching support is provided. The Chinese teacher G, in a step by step manner, builds a platform for students to understand the teaching process, such as asking questions about relevant knowledge, understanding the degree of students' knowledge mastery, and the depth of understanding. Students can naturally create a connection between the various elements of the problem, the subject of learning, and the situation of the problem. The Canadian teacher focuses on allowing students to try the problem, communicate and share their findings, and constantly explore and understand each step of the problem. Such understanding is often more profound, but it takes much more teaching time.

The fifth finding recognizes that most of the students can realize the transformation of speech representation and symbolic representation and can transform the image representation into speech representation and symbol representation. In these two lessons, the two teachers pay attention to the students' ability to transform their knowledge from one mode to another. The symbolic representation is transformed into speech characterization so that the student's ability to express can be trained in the classroom and whether the student understands the problem or not can also be evaluated.

From "solving the volume of irregular objects" lesson focused on the teaching of understanding mathematics problems taught by Chinese teacher G and the Canadian teacher C's "How many sticky notes can be filled with file cabinets?" lesson, the two classes involved four different levels of understanding. The study finds that each level of understanding has a certain meaning. The intuitive understanding and procedural understanding, which belong to instrumental understanding, can activate students' prior knowledge.

The abstract understanding and formal understanding, which belong to relational understanding as a high level of understanding, reflect the abstraction and logic of mathematics. Since both classes have mathematical problems that require abstract understanding, it is found that the mathematical problems of abstract understanding need to mobilize more knowledge and experience and produce more connections, which can promote the solving of mathematical problem using procedural understanding and formal understanding.

7 Suggestions

The purpose of teaching is to activate and support the internal learning process. Mathematical problem understanding teaching depends on the teaching process of solving mathematical problems. Therefore, teachers should play the role of a good guide and facilitator, build a platform for students' learning, and optimize the teaching design of mathematics problem solving.

7.1 *Teaching Problem Solving Conforms to Students' Cognitive Process*

Students' understanding of mathematics problems must go through the process of mathematics input, selective perception, problem coding, and problem representation. Teachers are the guides, collaborators, and supporters of students' learning. Students' learning requires the activation and support of teachers' teaching. Every stage of students' understanding of mathematical problems has its significance. When students are unable to understand the problem or have difficulty understanding the problem, the teacher needs to help clarify the understanding process of the student and provide support for understanding.

The first suggestion is to create an appropriate problem. The way the problem is presented and situation presented by the problem affects the understanding of problem. The problem presentation should be intuitive, clear, and in-line with the cognitive level of primary school students. The second suggestion is for teachers to mobilize the existing knowledge and experience needed. Selective perception determines which information becomes the main body of processing. Teachers can judge whether students understand the problem correctly according to their language description and extraction of main parts of the problem. Polya (1982) pointed out that, in teaching problem-solving, teachers need students to repeat the problem type, be able to explain the topic; and be able to point out the main parts of the topic, namely unknowns, known data, and conditions. Teachers should ask questions such as what is the known data? What are the conditions?

Students should consider the main part of the problem carefully, repetitively, and in all respects (Polya, 1982). "Finally, when students decode problems, teachers need to communicate with students to determine why students cannot decode smoothly. The reason may be that the students can not clarify the relationship between the conditions, or between the condition and the goal. Or the students can not store the extracted information into the working memory because of too much information and the complicated relationship.

7.2 *The Problem Should Be Presented in Accordance with the Internal Characteristics of Primary School Students*

Internal characterization is a cognitive structure used by students to understand problems. It is a series of operators to record and store information, so as to improve the structure of information. The influencing factors of the internal representation of the problem mainly include the following aspects: problem situation; the knowledge and experience of the problem solver; and the quality of thinking and other personality qualities of the problem solver.

Mathematical problems consist of words, charts, and symbols. In fact, the understanding of mathematical problems is also the understanding of the content presented by mathematical problems, that is, the information conveyed by mathematical words, charts, and symbols that can be processed. Among them, the mastery of verbal information is the key to help students obtain external representation as well as the tool needed for internal representation.

Problem presentation can directly affect students' problem understanding and solution. Problem presentation depends on problem situation. When problem situation can stimulate the existing schema in the mind, students can make a quick and positive response to the problem (Piaget, 1954; Zhou & Liu, 2014). Therefore, the creation of appropriate verbal information and problem situations can improve students' problem-solving ability. In teaching, teachers can help students to conduct internal representation of problems by creating situations that are related to students and presenting using language that students can understand.

7.3 *Activate Students' Existing Knowledge and Experience to Stimulate Decoding*

The mathematics learning of primary school students is the systematization of their common sense of life. Each student starts from their own real-life mathematical world and with that they learn to construct their own mathematical knowledge. Primary school students cannot learn mathematics without real-life experience. Students' prior knowledge and experience, including mathematical concepts, theories, methods, and strategies, are the basis for students to learn other new knowledge and understand problems. Understanding and solving mathematical problems require the learning of relevant mathematical concepts and theories. Effective learning is always carried out on the basis of experience.

Problem decoding is influenced by students' existing knowledge, experience, and memory. Therefore, mathematics teaching should pay attention to the connection between current problems and a student's prior knowledge and experience, so as to promote the meaningful processing and storage of information. Students' experience is mainly derived from their daily life experience and previously solved problems.

For students to solve problems, on the one hand, experience may help them understand the problem; on the other hand, it may hinder students' understanding, such as thinking stereotypes. Students will not be able to continue to understand problems correctly when their minds are trapped in past experiences that led to misconceptions. Therefore, in teaching, it is necessary to mobilize students' useful experience, establish meaningful connections between problems and known knowledge for processing, and finally store them in working memory for future problem solving.

References

Dai, Z. (1997). *Mathematical exercise theory* (p. 163). Shanghai: Shanghai Education Press.

English, L., & Gainsburg, J. (2016). Problem solving in a 21st century mathematics curriculum. In L. D. English & D. Kirshner (Eds.), *Handbook of international research in mathematics education* (pp. 313–335). New York, NY: Routledge.

Li, S. (2005). *Outline of modern teaching theory* (p. 141). Beijing: People's education press.

Liu, Y. (2007). Research on the characteristics of elementary school students' mathematical thinking. *Theory Circle, 9*, 129.

Ma, F. (2001). On two types of mathematical understanding. *Journal of mathematics education, 10*(3), 50–53.

Piaget, J. (1954). *The construction of reality in a child*. New York, NY: Basic Books.

Polya, G. (1982). *The discovery of mathematics: Understanding, research and explanation of problem solving*. Translated by Ou Yangjiang. China: Science Press.

Ministry of Education, PRC. (2012). *Mathematics curriculum standards for compulsory education* (2011 edition). Beijing: Beijing Normal University Press.

Shi, L. (2015). *Study* (p. 178). Beijing: People's Education Press.

Xu, S., & Connelly, F. M. (Project Directors). (2013). *Reciprocal learning in teacher education and school education between Canada and China*. Social Sciences and Humanities Research Council of Canada (SSHRC) 2013–2020 [Grant 895-2012-1011].

Zhang, W., & Fan, W. (2016). Educational mechanism and operational path of primary school students' mathematical language development. *Teacher Training for Primary and Secondary Schools, 11,* 47–48.

Zhou, W., & Liu, A. (2014). Research on the characteristics of primary school students' mathematical thinking. *China off-Campus Education, 3,* 103.

Zhong, Z. (2007). *Teaching theory of comprehension mathematics*. Nanjing: Nanjing Normal University.

Zhu, Z. (2009). *Child psychology* (p. 307). Beijing: People's Education Press.

CHAPTER 12

The Use of Manipulatives for Teaching Fractions in Two Canadian and Chinese Elementary Schools: A Comparative Research Analysis

Aihui Peng, Anthony N. Ezeife, and Bo Yu

1 Introduction

There is an important relationship between the use of manipulative materials and students' achievement in mathematics (Kira, Scott, & James, 2013). This indicates that manipulatives play a key role in young children's mathematics understanding and development. Piaget, Bruner, and Dienes have suggested that children learn by direct interaction with their

A. Peng (✉) · B. Yu
Southwest University, Chongqing, China
e-mail: 23166380@qq.com

B. Yu
e-mail: 1354701002@qq.com

A. N. Ezeife
University of Windsor, Windsor, ON, Canada
e-mail: aezeife@uwindsor.ca

S. C. Zhu et al. (eds.), *Reciprocal Learning for Cross-Cultural Mathematics Education*, Intercultural Reciprocal Learning in Chinese and Western Education,
https://doi.org/10.1007/978-3-030-56838-2_12

environment. Concrete materials provide a way for children to interact with mathematics concepts (Post, 1981).

Mathematics education is designated as social practices where the teaching and learning of mathematics actually occur, and it is deeply rooted in its particular culture (Peng & Nyroos, 2012). Canada and China are countries whose cultures are noticeably different. Our earlier study shows that Canadian and Chinese teachers use different manipulatives when teaching the same mathematical concept for the purpose of facilitating students' understanding (Peng, Ezeife, & Yu, 2018). In a Chinese school, for example, children learn how to regroup and count using sticks, while, in a Canadian school, students would use cubes or *Base Ten Blocks* to learn the same concepts (Peng et al., 2018).

Our current study further documents the differences in the use of manipulatives for teaching fractions in elementary schools in Canada and China. We focused on fractions based on the following considerations. Firstly, the concept of fractions is an important link to other mathematical concepts such as proportional reasoning, functional relationships, and measurement (Lamon, 2007). However, the teaching of fractions continues to pose a major challenge at the elementary and middle school levels. Secondly, research shows that, compared with the method that relies solely on the use of textbooks in teaching fractions, appropriate manipulatives help students learn fractions better (Sebesta & Martin, 2004). Thirdly, the topic 'fractions' is the teaching focus of our pair of research schools.

The following research questions were formulated to guide the study: What manipulatives are used to teach fractions in Canadian and Chinese elementary schools? Are these manipulatives different, and if so, why; and what are the implications?

2 Research Context

This study was conducted under a seven-year Reciprocal Learning Partnership Project in teacher education and school education between Canada and China (Xu & Connelly, 2014). It involves two Canadian and five Chinese universities, two Canadian School Boards, and over 40 Canadian and Chinese schools. The primary purpose of the project is to build a knowledge base for understanding and comparing educational views on Canadian and Chinese educational systems and for contributing to a knowledge-based public discussion of the reciprocal educational impacts of Canada and China. The project has five research teams—the mathematics education team, general education team, teacher education team, language and culture education, and science education team.

The mathematics team further created two intra-group teams of researchers and sister schools, which focus on selected areas of interest within the guiding mandate and goals defined by the mathematics research team. In this chapter, we report on the results from one of the two mathematics research teams arising from interactions with principals and mathematics teachers from a pair of sister schools.

For anonymity, the researchers use School A and School B to stand for the Canadian school and Chinese school, respectively. School A is located in Windsor, Ontario, Canada. It is a public school (Junior Kindergarten to Grade 8) with a student population of 300. Mathematics teachers in this school are generalists. School B is located in Chongqing, China. It is a public school with 3000 students ranging from Junior Kindergarten to Grade 6. There are currently 72 classes and 40 mathematics teachers who are specialists (Peng et al., 2018). These two schools have been sister schools since 2015. They hold monthly Skype meetings and embark on annual mutual visits, hence reciprocally learning from each other.

Every semester, some Skype meetings were dedicated to and focused specifically on problem solving. During these meetings, several topics that embrace problem solving such as mathematics games, computations, and fractions were discussed extensively. As participant researchers, we were fully involved in all the activities between the pair of sister schools.

3 Literature Review

A large number of studies have been conducted regarding the teaching and learning of fractions. In many studies, 'fractions' is considered to be a difficult topic for many students and teachers as many of the mathematical rules used for whole numbers operations do not apply. Lamon (2005) suggests that students may find that the meanings and models that they have been learning their whole life are not the same in fractions. Therefore, the use of manipulatives in teaching fractions is suggested as effective approaches to improve student understanding (Sherman & Bisanz, 2009). For instance, Bezuk and Cramer (1989) were emphatic about the use of manipulatives specifically for the teaching of fractions, stating that it "is crucial in developing students' understanding of fraction ideas" (p. 158).

Similarly, Kulm (1999) suggested that only adequately selected and relevant manipulatives would help students to develop a full understanding of the meaning of fractions. This is because such manipulative activities would expose the students to measuring, building models, using number lines, comparing/ordering fractions, and several other related fraction ideas easily demonstrable with the use of hands-on manipulative engagements.

The appropriate use of models has been helpful in the teaching and learning of fractions. The two key models, continuous and discrete, seem to be the most supportive in the teaching of fractions. The continuous model is well known to students as it refers to quantity or measure such as length, area, or volume (Behr & Post, 1992). This model also appears during the teaching of shapes and to address questions of "how much." The real-world applications include pizzas, pies, dinner plates, and some geometric drawings. The discrete model refers to a collection of separate objects. This model is helpful to students with questions about "how much." Common examples including counters, number of items in a package, and coins.

Recently, researchers have examined the efficacy of the use of discrete and continuous models and their impact on student understanding of fractions. Wilkerson et al. (2015) found that students understood fraction concepts better using discrete models. Other researchers investigated the use of different concrete and pictorial models on students' understanding of the part-whole construct of fractions and identified the strengths and limitations of the models used (Cramer & Wyberg, 2009). Their results show that pattern blocks did not contribute to understanding fractions. Paper fractions, on the other hand, were helpful in assisting students to order fractions with the same numerators but not in estimation activities. Dot-paper models were not helpful in teaching addition and subtraction of fractions with fifth grade students.

In summary, a number of studies have discussed the efficacy of manipulatives in mathematics education in a general sense (Larbi & Mavis, 2016; Moore, 2014; Ruzic & O'Conell, 2001). However, there is a paucity of information and lack of understanding about the use of manipulatives in teaching fractions in different cultural contexts. Our study was geared to making a contribution to this needed area of research from an international, reciprocal, and comparative perspective.

4 Research Methodology

4.1 Data Resources

Data were collected from interactions between the pair of sister schools including several Skype meeting notes, Canadian and Chinese mathematics textbooks, and video clips on teaching fractions from each school. The Skype meetings dwelt on the teaching and learning of fractions. In some of the sessions, researchers and teachers interacted and discussed extensively on the teaching of fractions. Subsequently, students from both sister schools solved sample problems designed and presented to them by mathematics teachers from the two schools. Each of the Skype meetings lasted one hour.

The Canadian mathematics textbooks offered in total four chapters on fractions from grades 3–6, one chapter in every grade, entitled Exploring Fractions, Fractions and Decimals, Addition and Subtraction of Fractions, and Multiplication and Division of Fractions. Each chapter on fractions began with an opening problem, followed by the themes Explore, Connect, and Practice, where various concrete manipulatives are suggested for illustrating fraction ideas. The Chinese mathematics textbooks introduced fractions only in the third and fifth grades and included two chapters, entitled Initial Understanding of Fractions and Meaning and Features of Fractions, while pictorial representations of different models are used to illustrate fraction ideas.

4.2 Analytical Framework

In order to understand manipulatives used to teach fractions in Canadian and Chinese elementary schools, we analyzed all of the data. In this chapter, we report the manipulatives presented in the textbooks since textbooks influence what teachers teach and how they teach, and we also include the results from video clips and Skype meeting notes.

In our study, a "manipulative" is an object that is designed in such a way that a learner can perceive some mathematical concepts by skillfully handling (manipulating) it. It may be physical/concrete, pictorial/static-visual, and virtual/dynamic-electronic. Furthermore, we use continuous and discrete models suggested by Behr and Post (1992) to categorize manipulatives. For the textbooks, we analyzed manipulatives suggested or presented under the title Lesson and subtitles Explore and Connect in the Canadian textbooks and example problems in the Chinese textbooks.

Since *Beginning Fractions* is the key to a deeper learning of fractions and the foundation for further computation involving fractions, our analysis mainly focused on *Beginning Fractions.*

Therefore, our analysis was conducted in grade 3 and grade 4 in the Canadian textbooks and in grades 3 and 5 in the Chinese textbooks. For the video clips, we analyzed the manipulatives used and recorded the time spent on using the manipulatives.

5 Findings

5.1 Manipulatives Presented to Teach Fractions in Canadian and Chinese Textbooks

Our findings show that manipulatives are widely presented in the Canadian mathematics textbooks. In the 16 lessons on *Beginning Fractions*, with the exception of the two lessons on Strategies Toolkit, concrete manipulatives were presented in the remaining 14 lessons. Furthermore, various concrete manipulatives are presented in the textbooks which include eleven types: pattern blocks, Cuisenaire rods, fraction strips, counters, stickers, paint tray, dimes, grid paper, color tiles, congruent squares, and base ten blocks. The most dominant manipulatives are pattern blocks, paper strips, grid paper, counters, and base ten blocks. Pictorial manipulatives are not widely presented in the Canadian textbooks. Among all of the manipulatives, the area model is the most popular model used in depicting fractions as it accounted for 57% usage, while the length and discrete models each had 21.5% usage, the two combining for a total of $21.5 \times 2 = 43\%$. Details are shown in Table 1.

Table 2 shows manipulatives used to teach fractions in the Chinese mathematics textbooks. From this table, we can see that pictorial manipulatives are widely used in the Chinese mathematics textbooks, whereas concrete manipulatives are far less used. In the 13 fraction problems where manipulatives are presented, pictorial manipulatives appeared in 11 problems, and there are only two concrete manipulatives that involve paper folding. Most of the pictorial manipulatives are geometrically shaped papers, and some are food pictures. In all of the manipulatives illustrated, the area model is the dominant one, accounting for 71% usage. The discrete and length models had 18 and 11% usage, respectively.

Table 1 Manipulatives presented to teach fractions in Canadian mathematics

Lesson	*Content*	*Manipulative*	*Model*
1	Equal parts	Pattern blocks, shaped picture	Concrete and Pictorial, Continuous (area)
2	Exploring fractions of a length	Rods, paper strips	Concrete, Continuous (length)
3	Exploring fractions of a set	Counters, stickers, paint tray	Concrete, Discrete
4	Finding fractions of a set	Counters, dimes	Concrete and Pictorial, Discrete
5	Naming and writing fractions	Grid paper, shaped picture	Concrete and Pictorial, Continuous (area)
6	Strategies Toolkit	None (word)	
7	Mixed numbers	Pattern blocks	Concrete, Continuous (area)
8	Fractions of a whole	Color tiles, congruent squares	Concrete, Continuous (area)
9	Fraction benchmarks	Paper strips	Concrete, Continuous (length)
10	Fractions of a set	Counter, fraction cards, dotted picture	Concrete and Pictorial, Discrete
11	Strategies Toolkit	None (word)	
12	Different names for fractions	Cuisenaire rods, paper strips	Concrete, Continuous (length)
13	More than one	Pattern blocks, grid paper	Concrete, Continuous (area)
14	Comparing and ordering fractions	Pattern blocks	Concrete, Continuous (area)
15	Exploring tenths	Base ten blocks	Concrete, Continuous (area)
16	Exploring hundredths	Base ten blocks	Concrete, Continuous (area)

5.2 *Manipulatives Used to Teach Fractions in Canadian and Chinese Classrooms*

The Canadian teacher prepared an 11-minute video clip about learning fractions. In the video, the students can be seen spending almost eight minutes exploring the manipulatives to better understand the fraction concepts or to solve problems involving fractions. The students used pattern blocks to show equal fractions and fraction strips to figure out all of the equal fractions. The Canadian teachers said that they hope "students learn abstract mathematics from concrete mathematics by letting them experience and operate manipulatives."

Table 2 Manipulatives presented to teach fractions in Chinese mathematics textbooks

Problem	*Content*	*Manipulative*	*Model*
1	Equal parts	Mooncake, shaped picture	Pictorial, Continuous (area)
2	Equal parts	Paper folding	Concrete, Continuous (area)
3	Equal parts	Mooncake, shaped picture	Pictorial, Continuous (area), Discrete
4	Equal parts	Paper folding	Concrete, Continuous (area)
5	Equal parts	Shaped picture	Pictorial, Continuous (length)
6	Equal parts	Shaped picture (coloring)	Pictorial, Continuous (area)
7	Simple computations on fractions	Shaped picture	Pictorial, Continuous (area)
8	Simple computations on fractions	Shaped picture	Pictorial, Continuous (area)
9	Simple computations on fractions	Shaped picture	Pictorial, Continuous (area)
10	Simple applications of fractions	Shaped picture (coloring)	Pictorial, Continuous (area)
11	Meaning of fraction	Candy, banana, cake	Pictorial, Continuous (area), Discrete
12	Fractions and division	Cake, mooncake	Pictorial, Continuous (length), Discrete
13	Fractions and division	None (word)	
14	Proper and improper fractions	Shaped picture (coloring)	Pictorial, Continuous (area)

In a 40-minute mathematics lesson provided by a Chinese teacher, manipulatives were used for only three minutes in the middle of the lesson when solving a fraction problem about using paper folding to show specific fractions. The Chinese teachers said that, although they recognized the variety of manipulative used by the Canadian teachers, they stated that manipulatives are mostly used in Kindergarten because "students should not rely too much on manipulatives, because they need to develop their abstract understanding of mathematical concepts through mathematical symbols."

6 Discussion and Conclusion

Our findings show that, in the Canadian school, various concrete manipulatives are widely used to teach fractions with pattern blocks, Cuisenaire rods, and fraction strips as the main ones, while pictorial manipulatives are far less used. However, in the Chinese school, pictorial manipulatives are widely used to teach fractions, while concrete manipulatives are used less often. The area model is the dominant model used in both schools. Furthermore, in the Canadian school, the length and discrete models are frequently used to the same extent, whereas in the Chinese school, both length and discrete models are used less often.

Our interview data show that the differences in the use of manipulatives for teaching fractions in China and Canada may be attributed to Chinese and Canadian teachers' different beliefs about mathematics in general. These beliefs are undoubtedly rooted in their distinct cultures and accordingly impact their teaching practices. Our comparative analysis gives insights for us to reflect on the implications for both sides: The Canadian school should be cautious that students not rely too heavily on concrete manipulatives as there may be need for them to gradually develop their formal knowledge of fractions; while in the Chinese school, students' mathematical ability may need to be taken into account, in such a way that the different fraction models are reordered from 'easy' to 'difficult', thereby enabling the optimal learning and grasping of fraction ideas.

Acknowledgement The researchers gratefully acknowledge the grant from the Social Sciences and Humanities Research Council (SSHRC) of Canada, which funds the larger Reciprocal Learning Canada-China Partnership Project (898-2021-1011, Xu & Connelly, 2019), and on which this mathematics-focused study is based. The National Social Science Foundation in China (17XMZ032), which also partially funded the study, is also hereby appreciated.

References

Behr, M., & Post, T. (1992). Teaching rational number and decimal concepts. In T. Post (Ed.), *Teaching mathematics in grades K-8: Research-based methods* (2nd ed., pp. 201–248). Boston, MA: Allyn and Bacon.

Bezuk, N., & Cramer, K. (1989). Teaching about fractions: What, when, and how? [Electronic version]. In P. Trafton (Ed.), *National council of teachers of mathematics 1989 yearbook: New directions for elementary school mathematics* (pp. 156–167). Reston, VA: National Council of Teachers of Mathematics.

Cramer, K., & Wyberg, T. (2009). Efficacy of different concrete models for teaching the part-whole construct for fractions. *Mathematical Thinking and Learning, 11*(4), 226–257.

Lamon, S. J. (2005). *Teaching fractions and ratios for understanding: Essential content knowledge and instructional strategies for teachers.* Mahwah, NJ: Lawrence Erlbaum Associates.

Lamon, S. J. (2007). Rational numbers and proportional reasoning: Toward a theoretical framework. In F. Lester (Ed.), *Second handbook of research on mathematics teaching and learning* (pp. 629–668). Charlotte, NC: Information Age Publishing.

Larbi, E., & Mavis, O. (2016). The use of manipulatives in mathematics education. *Journal of Education and Practice, 7*(36), 53–61.

Kira, J., Scott, M., & James, P. (2013). A meta-analysis of the efficacy of teaching mathematics with concrete manipulatives. *Journal of Educational Psychology, 105*(2), 380–400.

Kulm, G. (1999). Evaluating mathematics textbooks. *Basic Education*, *43*(9). Retrieved from http://www.project2061.org/publications/articales/articles/cbe.htm.

Moore, S. D. (2014). Why teach mathematics with manipulatives? *ETA hand2mind Research summary*, 1–4.

Peng, A., Ezeife, A. N., & Yu, B. (2018). Reciprocal learning in mathematics education: An interactive study between two Canadian and Chinese elementary schools. *Comparative and International Education, 47*(1), 1–14.

Peng, A., & Nyroos, M. (2012). Values in effective mathematics lessons in Sweden: What do they tell us? *The Mathematics Enthusiast*, *9*(3), 409–429.

Post, T. (1981). The Role of Manipulative Materials in the Learning of Mathematical Concepts. *Selected Issues in Mathematics Education* (pp. 109–131). Berkeley, CA: National Society for the Study of Education and National Council of Teachers of Mathematics, McCutchan Publishing Corporation.

Ruzic, R., & O'Conell, K. (2001). Manipulatives. *Enhancement literature review.* Retrieved from https://www.cast.org/ncac/Manipulatives1666.cfm.

Sebesta, L. M., & Martin, S. R. M. (2004). Fractions: Building a foundation with concrete manipulatives. *Illinois Schools Journal, 83*(2), 3–23.

Sherman, J., & Bisanz, J. (2009). Equivalence in symbolic and nonsymbolic contexts: Benefits of solving problems with manipulatives. *Journal of Educational Psychology, 101,* 88–100.

http://dx.doi.org/10.1037/a0013156 Wikipedia Encyclopaedia. (2018, September 21). *Manipulative (Mathematics education).* Retrieved from https://en.wikipedia.org/wiki/Manipulative_(mathematics_education).

Wilkerson, T., Cooper, S., Montgomery, M., Gupta, D., Mechell, S., Arterbury, K., et al. (2015). An investigation of fraction models in early elementary grades: A mixed-methods approach. *Journal of Research in Childhood Education, 29*(1), 1–25.

Xu, S., & Connelly, M. (2014). *Partnership grant reciprocal learning in teacher education and school education between Canada and China*. Retrieved from http://www.reciprocal-learning.ca/pages.php.

PART IV

Special Topics in Mathematics Education

CHAPTER 13

A Canadian Elementary Teacher's Differentiated Instruction in Mathematics Teaching: From the Perspective of a Chinese Researcher

Yingying Bai

1 Introduction

The learning needs of students in today's classrooms are very diverse. Some students are able to persevere through challenging tasks and block out distractions while others need constant positive reinforcement and a quiet learning environment. There are many ways to help students learn including scaffolding to understand a concept. There are also many ways that students learn including using visual or auditory cues. If teachers use a single teaching strategy in their classroom, they will invariably lose the interest of some of their students and could find that some of their students are not able to be successful (Karp & Voltz, 2000).

Y. Bai (✉)
School of Education, Central China Normal University, Wuhan, China
e-mail: baiying010@163.com

S. C. Zhu et al. (eds.), *Reciprocal Learning for Cross-Cultural Mathematics Education*, Intercultural Reciprocal Learning in Chinese and Western Education,
https://doi.org/10.1007/978-3-030-56838-2_13

The Equity Principle of the National Council of Teachers of Mathematics (NCTM), which emphasizes "high expectations and strong support for all students" (NCTM, 2000, p. 11), reminds us that teachers should cater for all students' learning needs to make sure that none of our students be isolated from mathematics learning. As such, teachers need to be mindful of the types of differentiated instruction that they use so that they can best support students' diverse learning needs.

The Ministry of Education in Ontario, Canada, has implemented curriculum and teaching reform that encourages "Support every child, reach every student". As teachers are the practitioners of the reformation of teaching, it is particularly important for teachers to cater to students' individual differences in classroom teaching. Through this case study of an elementary teacher in Toronto, we focus on how Canadian teachers vary teaching to accommodate students' diverse learning needs from the perspective of a Chinese researcher. The following two questions guide this study:

1. How do Canadian elementary teachers accommodate students' diverse learning needs in mathematics teaching?
2. How do Canadian elementary teachers view students' individual differences in mathematics learning?

2 Theoretical Framework

Differentiated instruction is largely based on the theories of social constructivism and multiple intelligences. The theoretical framework of mathematics differentiated instruction for this study was based on Vygotsky's (1978) zone of proximal development theory (ZPD), Gardner's (1983) Mutual Intelligence Theory (MI), and Tomlinson's theory of differentiated instruction (2008) with respect to students' diverse learning needs, and Gao's (2004) study of teaching conceptions.

According to Vygotsky (1978), each person has two stages of skill development. The first stage is the level the student can achieve by themselves. The second stage is the level they can achieve with a mentor or teacher. The distance between these two levels of actual and potential development is called the "zone of proximal development" (ZPD). Accordingly, the teacher must provide students with assistance at a level beyond independent learning yet within their ZPD in order for the child to learn new skills (Bruner, 1981; Vygotsky, 1978).

Gardner's multiple intelligence (MI) theory provided a model for how student learning differences might be regarded (Armstrong, 2009; Gardner, 1983). Instead of seeing students as either intelligent or unintelligent, MI Theory allows teachers a framework for seeing the potential in every student (Gardner, 1983). It is generally accepted among educators as an instructional tool for understanding learner variation and an integral part of using differentiated instruction (Dodge, 2005; Voltz, Sims, & Nelson, 2010).

Tomlinson's (2001) model of differentiated instruction is based on the belief that students learn more effectively when teachers accommodate instruction in response to their learning differences and similarities. For Tomlinson (2004), the basis of differentiated instruction is that all students are different and therefore should be taught in such a way where diverse learning differences are addressed. This means that the content, process, and product of instructional design can all be differentiated (Tomlinson, 2001).

In mathematics teaching, differentiating the content, process, and product is a relational process that builds off of one another, and is, as a whole, cycled through learning environment continuously, as each student grows and develops in their learning capacity. Thus, the theoretical framework of this research is based on four components of mathematics differentiated instruction, and six dimensions of teaching conceptions.

3 Methods and Data Sources

This study adopted the orientation of the qualitative research, and the case study methodology was used as the research strategy (Yin, 2002). Our goal was to find the teachers' conceptions and practices of teaching while accommodating diverse learning needs in mathematics teaching.

Our study was conducted from October 2017 to May 2018. We gathered data from three sources: classroom observations, interviews, and the relevant data of the school curriculum and the teaching practices. These data were collected from teacher Kelly from Pleasant Hill Public School (names are pseudonyms) during the study.

Kelly was observed 17 times (two blocks each time) as she taught her mathematics class. During the classroom observations, field notes were taken immediately about what was observed and some potential follow-up questions according to the observation. Most of the short interviews about lessons were further discussed with Kelly in her prep or recess time. The semi-structured formal interviews included questions about the teacher's conceptions of teaching and accommodations of students' diverse learning needs and differentiated instruction. Most of these interviews were conducted after class depending on Kelly's schedule. We also interviewed one principal and one coach at this school. All interviews were audio-taped and transcribed, and the participants had the opportunity to review transcripts of their interviews.

Pseudonyms for the teacher and the school were used in this study to ensure confidentiality. Specific details about the school have been omitted. Qualitative analysis methods were used to organize the data and to conduct analysis to uncover patterns in the data. We used a constant comparative method. Earlier data were analyzed soon after data collection to decide upon core themes that would be further investigated in the remaining study (Glaser & Strauss, 1967).

4 The Case of Kelly

The overall workflow for this study is observing the teacher's teaching practices first and then conducting interviews and more follow-up questions with the teacher based on the discovery from the observations. The goal is to reveal the teacher's conceptions of teaching from the teacher's description in catering to students' individual differences in the teaching of mathematics.

Kelly has 15 years of teaching experience. Her first job was in a company. She took the TOEFL exam after four years of work, and then she went to an Asian country to teach adult English for one year. It was at that time that she found she loved teaching so much. When she came back to Canada, she took the education courses at the Teachers' College, and finally she became an elementary teacher. Kelly has 20 students in her grade 3 class this year. She taught almost all the subjects except the gym.

4.1 *Kelly's Practices of Teaching in Catering to Students' Diverse Learning Needs*

4.1.1 *The Content of Teaching*

Kelly usually used open-ended questions in mathematics teaching. In the design of open-ended questions, two strategies were used frequently by Kelly: one is encouraging students to solve problems in a variety of ways, and the other is enriching the choices of tasks, so that students can create their own questions based on their understanding. Sometimes, Kelly would choose one question designed by the students as the task for the whole class, which is also a good way to engage students.

Kelly thought that an open-ended question is a good way for teachers to know whether the students understand the key concepts clearly, and do students can transfer the knowledge flexibly when it comes to different situations. Depending on the learning needs and the difficulty of the questions, students were usually encouraged to solve 1–3 open-ended questions in two blocks.

To illustrate Kelly's use of open-ended questions, I will describe one of the lessons of "Probability" that I observed. In the first 19 minutes of class, Kelly first explained the tasks to students, which was to pick an object out of a brown paper bag. She then demonstrated the process and then had her students completed 10 extractions and record the results; the process led the students to recall different scores and asked students to discuss how to represent the data in a better way. Kelly introduced that activity:

> **Kelly**: "Ladies and gentlemen, you are going to have a choice, you can work by yourself or with a partner. You are going to record your results and you are going to predict which bag you mostly have."
>
> *The activity is there are 5 paper bags numbered ABCDE hung on the blackboard. Each bag has 20 color tiles, including four colors of blue, red, yellow, and green. However, the number of color tiles in each bag is uneven. The number composition of each color is written on the bag. And two boxes are placed in the center of the carpet in the classroom. There are secret bags prepared by Kelly according to the number of students, which are for students to do activities. Students were asked not to open the secret bag to view the composition of color tiles, but to follow the criteria on the poster.*

> *The criteria of the task are written on a poster, which are:*
> *1. Take a color tile out of the bag.*
> *2. Record your results.*
> *3. Repeat 10 times.*
> *4. Represent the data in a way that make it easier to read and understand.*
> *5. Describe the data using fraction words.*
> *6. Predict which bag you mostly have and explain your thinking.*
>
> **Students**: *After the students report to Kelly whether or not to select a group to work with and confirm the companion, they pick up the unopened secret bag from Kelly and begin to record and make the data predictions.* (Kelly observation, April 19, 2018)

Kelly argues that it is very important for teachers to understand the big ideas in the mathematics curriculum, so that different strands can be integrated with each other and teachers can flexibly design activities rather than mechanically executing every content of the mathematics curriculum in sequence. Instead, Kelly thought that teachers should think about what kinds of questions can achieve the teaching of these big ideas, and then connect the big ideas with the students' real lives and group activities, to design truly interesting and meaningful activities to engage students in math learning. Kelly said:

> I think, stop using the textbook, stop using the textbook, because the textbook doesn't make you love math, love math do math think about the math think about what the kids are going to do, and then let the kids surprise you, right?…… I think teachers need to do the same things that kids do, they need to talk about the math, they need to learn more math they need to share with others, they need to use the tools so they know how to use the tools. Everything that kids do teachers should do too, but I think of teachers love it, the kids will love it. (Kelly interview, March 7, 2018)

4.1.2 The Process of Teaching

Kelly believes that teachers should pay attention to each student's zone of proximal development. She usually adopts different strategies for students of different learning needs. The strategy she usually uses was to change the numbers in the question to adjust the difficulty of the problem according to the level of the students. Kelly believes that differentiated instruction in mathematics is that every student can do the math, and every student can explore mathematics on the basis of his or her ability.

Kelly also focuses on problem-based learning by providing individual guidance to students, asking students questions constantly, and helping students to understand concepts and to find solutions step by step. The following is a conversation between Kelly and a student during a lesson on different ways to represent one-fourth (T represents Kelly and S represents Student):

T : Does this show one-fourth?
S : No...
T : What can you do, what can you add from what is in my hand, to make sure that this one shows one-fourth. [*Kelly held two equilateral triangle color tiles and two parallelograms color tiles in her hand.*]
S : [*The student took a triangular color tile from Kelly's hand and put it into the graphic. He tried to put the new graphic into a regular shape for several times, and finally, he forms a graphic composed of two parallelogram color tiles and two triangular color tiles.*]
T : What did you do? You didn't change anything.
S : [*The student changed the layout of the graphic again and placed it into a graph consisting of three parallelogram models.*]
T : How many equal parts do you have here? [*Kelly asked the student about the new graphic.*]
S : ... [*Student silence.*]
T : Do you have other parts that equal to this one? [*Kelly pick up a parallelogram color tile from the graphic and asked the student.*] Which one is equal to this one?
S : This one... [*The student pointed to a parallelogram color tile when he said.*]
T : And this one is equal to this one? [*Kelly placed one parallelogram color tile on the parallelogram graphic which composed of two equilateral triangles and asked the student.*]
S : Yeah...
T : So how many equal parts do you have? One, two... How many? [*Kelly asked by pointed the graphic*]
S : Four...
T : How many equal parts?
S : Two...
T : No, no, no.
S : One.

T : Don't guess! Look, you guess you give me numbers. How many equal parts? [*Kelly asked by pointed the graphic.*]
S : Three… [*The student said with a small voice.*]
T : [*Kelly show it to the student by counted the color tiles.*] Three! One, two, three. Take something from my hand, make four equal parts. [*Kelly held two parallelogram color tiles and two triangle color tiles for the student to choose.*]
S : [*The student took a triangle from Kelly's hand and tried to add it to the original graphic, but it seems that he did not tend to change the original hexagon.*]
T : You are stuck with the way what it looks now, it's going to look different.
S : [T*he student changed the layout of the original graphics and continued to change shapes.*]
T : It must look different… Is that enough? Now, do you have four equal parts?
S : I think…no…
T : What else do you need? [*Kelly extended the two parallelograms color tiles and a triangle color tile in her hand to the student.*]
S : [*The student took out a triangle color tile from Kelly's hand and added it to the graph to make an adjustment, which is consisting of four equal parts, four parallelograms.*]
T : Good job! I'm so happy you figure it out all by yourself. (*The product created by the student before and after Kelly's guidance is shown in* Fig. 1.) (Kelly observation, April 5, 2018)

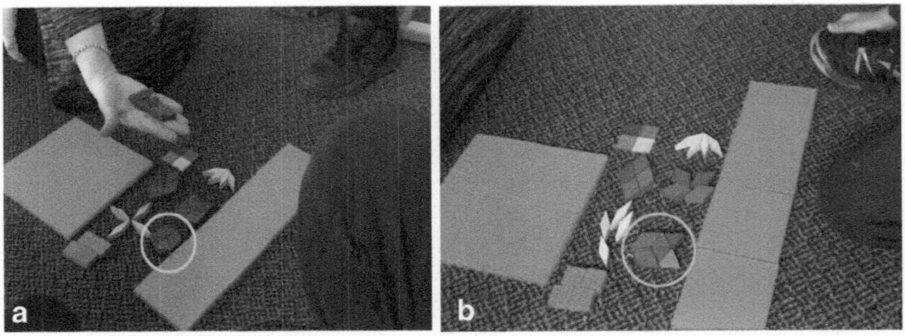

Fig. 1 Kelly guides one student through a misconception in their mathematics inquiry. **a** Student's product (in the circle) before Kelly's guidance, **b** Student's product (in the circle) after Kelly's guidance

4.1.3 The Product of Teaching

Kelly believes that it is important for students to be able to discuss math, express their ideas clearly, and visualize the ideas with manipulatives, which also is an important way for teachers to diagnose whether students understand what they are learning. After the students complete the task, Kelly usually asks them to write their thoughts and practices on the notepaper, and then Kelly posted the notes on a poster. After all of the students have finished writing their thoughts, the students discuss their notes with each other.

She often encourages students to speak out wherever and whenever they do not understand. For hands-on tasks, Kelly typically uses an iPad to take pictures of the students' product and project it onto a projection screen to discuss it with the students after group assignment. For assessment, the most common strategy used by Kelly is to design open-ended questions and change the numbers in the question to examine students' levels. Kelly states:

> the way I typically do is I give them a problem to solve and I see how they solve the problem and then I see, I see what do they know and what do they not know, what do they almost know, so that kind of idea. What do they already have that I do not have to teach them again. What are they so far away from understanding that I really need to do a lot of teaching for....... I think for me the biggest thing I need to know is what numbers can they manage and what strategies do they have because then we have someplace to start. The only way for me to do that is by asking them problem and having them figure it out. (Kelly interview, April 5, 2018)

4.1.4 The Learning Environment

Kelly believes that building a good learning environment for mathematics requires students to understand the learning in the community, and teachers must establish a learning community in the classroom. It is important that students feel safe and know that everyone needs to learn and work together because everyone has their own advantages and need to respect each other so that students feel safe and confident in the classroom. Kelly emphasizes:

> I wanted them to see that, in our community, we have a lot of different people who see things in different ways, and just like we see it in different ways we learn different ways, so we have to work together and accept each other and try things in different ways. (Kelly interview, March 6, 2018)

Furthermore, Kelly states:

> you have all the tools that kids need to use it. You have the materials, kids need to know how to find it. They don't need your permission to use it...... you need space in your classroom. And you should not be afraid to use the hallway too. Because sometimes kids need to spread out. And if they're using different kinds of materials, you might want to spread the materials out too. So I think you need space in the room and they need to see their work on the walls. (Kelly interview, April 19, 2018)

4.2 *Conceptions of Teaching in Catering to Students' Diverse Learning Needs*

4.2.1 *The Conceptions of the Nature of Mathematics Learning and Students' Roles*

Kelly believes that there are a lot of differences in students' prior knowledge when it comes to mathematics. These differences include students' learning experiences and family background. Many students cannot use numbers flexibly because of the lack of number sense. She firmly believes that there are some individual differences between students and that every student can achieve his or her potential within their zone of proximal development. Kelly states that:

> I think their prominent primary knowledge is lacking, but I think what's lacking is that they do not have flexibility with number. I think as a grade one teacher and a grade two teacher, I think that's the crucial thing is to teach kids to be flexible with numbers, to understand number relationship, because that is the key to understanding everything that we're doing in math. And there's a big jump between grade one and two and grade three, because in grade three you're junior, so in grade three you're supposed to know basic math, you're supposed to be able to read and understand what you read, it's a huge difference. I think that I think that a lot of kids they need to have that number sense, but they don't have it. (Kelly interview, March 7, 2018)

Furthermore, Kelly believes that:

> if they are not confident with number, they don't really see the relationship between numbers...... the more that they can see the relationship between

> the numbers and how they're similar and different, then the better they are going to get with math. (Kelly interview, March 5, 2018)

Kelly believes that understanding and using numbers fluently are critical to students' mathematical learning. Students need to learn to read fluently, choose numbers, recognize the relationship between numbers, and use numbers to solve problems freely.

4.2.2 *The Conceptions of the Nature of Mathematics Teaching and Teachers' Roles*

When we asked "What do you think is the most important thing for teachers' teaching in mathematics?" Kelly says that she puts great emphasis on a teachers' love of mathematics and a teachers' flexibility in mathematics teaching. Kelly believes that sharing ideas, using manipulatives, working with others, giving students tasks they need to work hard at but are capable of accomplishing, and designing tasks based on students' prior knowledge are five very important aspects of teaching that can engage students effectually. Kelly states:

> So being able to share their ideas, being able to use tools and being able to work with others and having work that they can manage on their own with a little bit of struggle. I think… and work that builds on what they already know. So I think those five things will engage kids, because I think, I think if they can do the work struggle a little bit and still get it, they're going to want to do it, if they can talk with someone else and share their ideas or listen to the other person's ideas, then they're going to feel, 'Oh I know what to do!', 'Oh thanks, I can get that.', or 'How did you do that?'. I think that it will help them to be engaged, and then sharing their learning sharing their thinking, I think that will help them to be engaged as well. (Kelly interview, March 7, 2018)

Kelly emphasizes the important role of manipulatives in helping students to understand mathematical concepts and solve problems. Kelly states:

> I think a big factor is kids need material, so they need manipulative, and I think some teachers, not intentionally maybe, but some teachers make kids believe that if you using the math tools, it means you're not strong in math, and I think that if kids believe that they're not going to use the tools, because they are going to think that means that they are dumb, that

> they need the tools and nobody else does. So I think teachers need to stop giving kids that idea, I think using tools helps, you have to choose the right tool, and use the tool in the right way if you need to, and is whether you use it or not is a choice, it does not matter. (Kelly interview, March 7, 2018)

Kelly firmly believes that how teachers teach is the most important factor affecting students' mathematics learning; teachers should pay attention to students' preparation level of learning, learning style, and interest in mathematics teaching; and teachers should learn with students with a growth mindset.

4.2.3 The Conceptions of the Aims and Expected Outcomes of Teaching

For the question "What progress do you expect students to make in mathematics in this semester?" Kelly said that she hopes that students can see the relationship between numbers and understand the relationship between the value of the coin and the quantity, and linked it with different strands such as fraction, pattern, data management, and equations, and students learn to look at questions from a mathematical perspective and understand that everything in mathematics is interrelated. She further explains, stating:

> It is about them being more flexible with numbers, so as we do money I want them to remember that when we did patterning we count by the same amount, so I want them to be able to, being able to fluently count money, so pennies are counting by one, the nickels are counting by five, dimes by ten and then quarters by twenty-five, but groups of four. Then I want them to be, to tie money into fractions, why is a quarter called a quarter, because it is a quarter of a dollar, so what is the dime, a dime is a tenth of dollars, so I want them to see to be flexible with their thinking, I want them to see how the number of coins in the bag, is different than the value of the coins in the bag...... I want them to be able to see the relationship in the connection like a nickel is half of a dime, but also a fifth of a quarter, so I want them to see, so we get a time money into fractions bring back patterning, maybe tied back into data management, and we can do it with quality of equations as well, right? We can tie into multiplication again, and division, and because we're doing the fraction. So that's what I want, I want them to see that everything in math is connected, and I wanted to start looking at it with math eyes, so I wanted to see the relationship between the numbers and why did we choose these values for

> the coins, because I want them to be able to understand that there is no three-cent coin, why? I want them to start thinking about these kinds of things as well, there's no seven-cent coin, why? So those kinds of things. (Kelly interview, March 7, 2018)

Kelly indicated that learning expectations should consider students' learning abilities and levels, and attach importance to the cultivation of mathematics application ability.

4.2.4 The Conceptions of the Content of Teaching

Kelly believes that teachers need to understand the big ideas in mathematics curriculum and design learning tasks through open-ended questions flexibly, and students should master the multiple relationships between numbers and learn to look at the numbers from different ways. Kelly states:

> I think, teachers need to think about what are the big ideas in the curriculum, and they need to think of questions that will meet those big ideas and questions that are challenging and inquiry-based, open-ended, so everybody can answer it... I think the thing that would help teachers to differentiate for every student is to stop looking at the curriculum like a checklist, 'I have to do this I have to do that...' no, look at the big idea and how can we, what questions can we ask that will answer this big idea, that will work on this big idea, but that's engaging, and if we can, can we make it related to their life or their community. (Kelly interview, March 21, 2018)

Kelly believes that the multiple relationships between numbers are very important for students to understand the quantitative relationship. Students should learn to count and understand the relationship between numbers, which requires teachers to create opportunities for students to look at numbers in different ways. She further explains, stating:

> I think the strategies that help them with the math is, understanding doubles, so that's very important, understanding doubles and being able to count forwards and backwards easily, and being able to count by a variety of different numbers, and also being able to... so the kids that do the best are the kids that can see that two times three is six so twenty times three is sixty, the kids who can see that and understand that if you multiply by ten it's ten times, greater. So I think that the kids who say

> just at a zero, they don't understand that it's getting ten times greater, so I think the strategies that work for them are strategies that help them to look at numbers and understand the relationship between numbers……So the strategies that help kids are help them look at numbers in different ways……when kids start and look at numbers fluently, then they would be successful. So just strategies like that like choosing the right number, giving them lots of opportunities to be to see numbers in different ways. (Kelly interview, March 7, 2018)

4.2.5 *The Conceptions in the Methods of Teaching*

In the methods of mathematics teaching, Kelly advocates encouraging students not to be afraid of making mistakes. Continuous positive attempts are the premise of learning, and letting students learn to explain and share their ideas is also an important way to promote learning. Collaboration, sharing, and building a learning community are critical in building a safe learning environment. Kelly pays close attention to building students' self-confidence in learning mathematics, and she advocates that students' self-confidence in learning mathematics should be built at the beginning of mathematics learning. She states:

> We want the kids to be confident with making mistakes, and we also want kids to understand what is important when you're working as a student when you're learning. The five keywords are 'Effort', 'Risks', 'Success', 'Perseverance', and 'Mistakes'. So we want kids to understand that there is learning does not happen unless you make a mistake…… when you make a mistake your brain is firing you're learning you're making connection, if you don't make a mistake your brain is not growing, so if I asked everybody in the class what's one plus one, they all know it, yeah you got it all right, but did you learn anything? No. So I want them to feel comfortable with making a mistake. (Kelly interview, March 6, 2018)

> So even as teachers we have to get away from just focusing on intelligence, and everybody can be creative…… and smart children do have learning problems and parents should not protect their kids from failing, because if you don't fail you don't make a mistake you do not learn. (Kelly interview, March 6, 2018)

Kelly said that listening to students' ideas and encouraging students to express and explain their ideas is an important way for teachers to understand and help students. Therefore, she advocates that teachers should encourage students to ask questions, express, and explain their ideas to promote mathematical understanding. She further explains, stating:

> Every idea they have, they have for a reason, so even if it's not the idea I would have, but it might be a really cool idea. Like I think that if they are allowed to share their ideas with each other or with me... So Ju** for example, she's making great progress but at the beginning of the year she was making no progress she was going down, but we had a conversation and I spoke to her and I said you need to share your ideas with me, you need to ask questions and because that will help me to teach you better, and she asked me, 'But how does us asking you questions help you to teach us?', and I said to her, 'Because then I know what you're thinking and maybe you're thinking something that's so incredibly awesome that I never thought of, then I can help you and I can understand what you're thinking, and I can help you get to the other place.' Or maybe I said, 'Maybe you're thinking something that is interesting but it's going a little bit away from what we need to be doing, and I will know that because you're telling me your thinking, and then I can help you bring it back, so you can see how your idea can be extended and connected, so you can solve the problem.' It was just like all for her, so now, she will see on the carpet she sits at the front, and she asked me 'I don't understand that, can you explain that? That doesn't make sense.' Now she does it, all the time, and she is making excellent progress!...... so being able to explain your thinking or even to share your thinking, it makes a lot of progress. (Kelly interview, March 7, 2018)

In addition, Kelly thought that it is very important to establish a learning community in mathematics learning. It is necessary for students to understand that each person has his own abilities, advantages, and areas for improvement. Learning needs cooperation and sharing. The active communication, sharing, questioning, and discussion between students are very important. She states:

> It has to be a community of learners. They have to see that we all work together. We all have strengths and areas that we need to improve on, I think the learning environment for math needs to be one... you have to build community in the classroom. I think it's important to let kids know you're not gonna like everyone, but you have to be respectful to everyone. You have to listen to their ideas. Be willing to share your ideas. The best learning environment for math is where children are working in different fluid groups and they are sharing their ideas and building on each other's

> ideas and questioning each other. 'You said this, I don't understand that. Can you explain that to me?' Or 'I disagree with that, I think this.' …… I think the best community for mathematics learning is a community where everybody is… everybody sees each other as important to be able to do the work…… And then everyone has something to share and you build the confidence where you can talk to the other person. You can listen and you can be wrong. And it's okay to be wrong. I think the best environment for math is where a kid can say 'Wow! My answer is completely wrong. I really love the way you did that! Can you show me that again?' Or 'I don't understand that. Can you show me that again?' Or 'Can you show me that in a different way?' I think, I think as a teacher when I hear that, that's just like heaven. (Kelly interview, April 19, 2018)

4.2.6 The Conceptions in the Assessments of Learning

In assessment, Kelly said that she often uses two strategies. First, at the beginning and end of the unit, give the students the same assessment to test what the students have learned, let the students see their progress and reflect on their learning. Kelly explains:

> I look at their work, and sometimes I give them a diagnostic assessment at the beginning of the unit, then we do some work and then they give them the same diagnostic assessment at the end of the unit, and I let them see how they're doing on it. I gave them the diagnostic, then we did a lot of learning and then I gave them the same test again after we were finished after we did a whole lot of learning. And then I highlight the parts that were incorrect and gave them an opportunity to fix it, then I stapled them together, so the kids could see the progress that they made, then I had the kids write on posted notes, what goals they have what do they, so what did you learn about your learning, so what did you improve on and what goals do you have to become even better, so that's one way that I assess. (Kelly interview, March 7, 2018)

Another strategy used by Kelly is to observe the students' learning during the process of teaching and student group activities and then group them for guidance. Kelly explains:

> Other ways it's just marking the work and deciding who understands like I do for language, group in the kids. So maybe this group needs help, so we're doing money. So maybe if I find that group doesn't understand the coins and their values, then I can divide a problem for them and then I can have a different problem for kids maybe they understand the values

> but they're not sure how to add them, so then this group will be adding money. And then maybe this group knows how to add money so they're learning about making change, so it's so doing stuff like that, so look at their work and then decide what they need. (Kelly interview, March 7, 2018)

> Usually I will give them a math problem, that helps me see if they understood what we were doing in the unit, so that if they understand, then they have it. I don't like the idea of testing kids all the time, because then they feel like... because I don't want kids to think that it's all about the answer, it's about the process. And if I give quizzes and tests and exams then it's giving them the idea that it's all about the answer, and I don't want it to be about the answer, the answer is important, the answer you need strategies and you use strategy so you can get the correct answer, the answer isn't important but until they are comfortable with strategies to get the answer, we need to focus on the strategies. (Kelly interview, March 7, 2018)

5 Conclusions and Discussions

Analysis of Kelly's mathematics teaching practices revealed that designed parallel tasks and open-ended questions are the most common strategies used by her to enable all students to have the opportunities to participate in mathematics learning. Another characteristic of Kelly's teaching is the mathematics tasks in her class are usually based on real-life situations, and she will vary the tasks flexibly according to the students' level of preparation, learning style, and the way of learning. The importance of manipulatives was emphasized a lot to help students understand mathematical concepts and visualizing their thinking. Students were encouraged to use a variety of ways to express their learning products, and Kelly used observation, communication, discussion, and other ways to conduct an ongoing formative assessment of students throughout the teaching process.

In addition, Kelly tended to pay attention to the arrangement of the physical learning environment in order to help students to become independent learners, and she also focused on creating a safe learning environment for students to feel free to take risks and make mistakes.

In terms of Kelly teaching philosophy, we found that she firmly believes that there are some individual differences between students and that every student can achieve his or her potential within their zone of proximal development. She believes that how teachers teach is the most important factor affecting students' mathematics learning. From her perspective, teachers should pay attention to the way they teach mathematics, and how they cater to students' interests and diverse learning needs. Furthermore, teachers should learn about mathematics with their students, keeping a growth mindset as they learn.

Kelly recommended teachers be flexible with mathematics curriculum and integrated different strands in the curriculum to ensure that most of the big ideas are covered. Kelly hopes that students can find mathematics everywhere. Students are expected to be good at discovering mathematics in life. It is advocated that teachers should take the student preparation level as the starting point, pay attention to learning interest and learning style to design mathematics tasks. It is considered that the assessment is to prepare for teaching.

There are a number of factors affecting Kelly's differentiated instruction, which include the personal factors, the factors from school, and also the factors out of school. The personal factors that this study found include teachers' beliefs and values of teaching, the teaching experience, and their pedagogical knowledge and ability. Some personal factors like teachers' personality, personal experiences, and teachers' learning and self-reflection, and other personal characters also affect their practices and conceptions in catering for students' individual difference in mathematics teaching. The beliefs of teaching, the teaching experience, and their personal experience determine how the teachers think about students' mathematics learning and the nature of mathematics teaching.

The teacher's pedagogical knowledge is the ability of the teachers to cater to students' individual differences. The teacher's learning and self-reflection are the endogenous power to promote the interaction of other factors. Kelly said that she is quite confident in teaching mathematics and that the ability to take care of students' diverse learning significantly improves as teaching experience grow. Kelly believes after she taught almost all grades, her ability to cater to diverse learning drastically improved.

At the same time, how teachers cater to students' individual differences is also influenced by the factors from school that include the school culture, the principals, the organized professional activities in the school, colleagues, and students. The school culture affects teachers' conceptions and practices to students' diversity. The principal is a very important person who influences the professional development of teachers. Kelly is very grateful to the principal for his support in teaching resources and professional development programs. Kelly said that collaboration and discussion with colleagues, especially the opportunity to co-plan with colleagues twice or three times a week, provide her more strategies and new ideas to cater to different students' needs. Kelly said she also benefits from the organized professional development activities like Professional Learning Community meetings in school. There is a certain degree of interaction between the school situation and teachers' practices in catering to students' individual differences.

In addition, the educational policy of Ontario organized professional activities out of school, and the support from important people out of school are other factors that have a certain impact on the teachers' catering for students' individual differences. Ontario has a series of pre-service and in-service policies to promote the professional development of teachers, and also developed some teaching documents to stimulate teachers catering to student individual differences and differentiated instruction.

Kelly said that taking Additional Qualification courses and attending workshops in the summer are also more influential in promoting the professionalism of teachers catering to students' diversity. Some mathematics experts like Marian Small and Jo Boaler also play an important role in influencing teachers' conceptions and strategies in differentiated instruction.

Ontario is Canada's largest province and has nearly two million K-12 students, which accounts for about two-fifths of Canada's K-12 students. "In terms of diversity, 27% of Ontario students were born outside of Canada and 20% are visible minorities. Toronto, the main city in Ontario, is one of the most diverse cities in the world" (OECD, 2011, p. 71). The diverse demographic composition promotes the collision and integration of multiculturalism in Ontario and provides a rich experience in meeting the diverse learning needs of students in K-12 education. This research on the needs of students' diverse learning needs in elementary mathematics education in Ontario, Canada, can provide useful reflection for China's elementary mathematics curriculum and teaching reform in paying attention to students' individual differences and respecting the learner diversity.

References

Armstrong, T. (2009). *Multiple intelligences in the classroom* (3rd ed.). Alexandria: Association of Supervision and Curriculum Development.

Bruner, J. (1981). The social context of language acquisition. *Language & Communication, 1*(2), 155–178.

Dodge, J. (2005). *Differentiation in action.* New York: Scholastic.

Gao, L. B. (2004). *A study of Chinese teachers' conceptions of teaching.* Wuhan: Hubei Education Press.

Gardner, H. (1983). *Frames of mind: The theory of multiple intelligences.* New York: Basic Books.

Glaser, B. G., & Strauss, A. L. (1967). *The discovery of grounded theory.* Chicago: Aldine.

Karp, K. S., & Voltz, D. L. (2000). Weaving mathematical instructional strategies into inclusive settings. *Intervention in School and Clinic, 35*(4), 206–215.

National Council of Teachers of Mathematics (NCTM). (2000). *Principles and standards for school mathematics.* Reston, VA: NCTM.

OECD. (2011). *Strong performers and successful reformers in education: Lessons from PISA for the United States.* Paris: OECD Publishing.

Tomlinson, C. (2001). *How to differentiate instruction in mixed-ability classrooms* (2nd ed.). Alexandria: Association for Supervision and Curriculum Development.

Tomlinson, C. (2004). Differentiation in diverse settings: A consultant's experience in diverse settings. *The School Administrator, 7*(61), 28–35.

Tomlinson, C., Brimjoin, K., & Narvaez, L. (2008). *The differentiated school: Making revolutionary changes in teaching.* Alexandria, VA: Association for Supervision and Curriculum Development.

Voltz, D., Sims, M., & Nelson, B. (2010). *Connecting students, teachers, and standards: Strategies for success in diverse and inclusive classrooms.* Alexandria, VA: Association for Supervision and Curriculum Development.

Vygotsky, L. (1978). *Mind in society: The development of higher psychological processes.* Cambridge, MA: Harvard University Press.

Yin, R. K. (2002). *Case study research: Design and methods* (3rd ed.). Thousand Oaks, CA: Sage.

CHAPTER 14

Meaningful and Sustainable Mathematics Education for Students of Indigenous and Ethnic Minority Cultural Backgrounds in Canada and China

Anthony N. Ezeife

1 Identification

Who are Indigenous Peoples?

> In its identification of *Indigenous Peoples*, the Wikipedia Encyclopedia (2018), states:
>
> Indigenous peoples, also known as first peoples, aboriginal peoples or native peoples, are ethnic groups who are the original inhabitants of a given region, in contrast to groups that have settled, occupied or colonized the area more recently. Groups are usually described as indigenous when they maintain traditions or other aspects of an early culture that is associated with a given region. (p. 1)

A. N. Ezeife (✉)
University of Windsor, Windsor, ON, Canada
e-mail: aezeife@uwindsor.ca

S. C. Zhu et al. (eds.), *Reciprocal Learning for Cross-Cultural Mathematics Education*, Intercultural Reciprocal Learning in Chinese and Western Education,
https://doi.org/10.1007/978-3-030-56838-2_14

Along the same line, the term "Indigenous" is defined as "native to a particular country, region, etc., not brought from elsewhere" (De Wolf, Gregg, Harris, & Scargill , 1998, p. 780). In general, by inference, *Indigenous Peoples* refers to culturally distinct groups who have been affected or influenced by colonization (Hitchcock & Vinding, 2004). Thus, the term "Indigenous populations" is used in this chapter to refer to those people who were the original natives of a given country or geographical region before contact, and often permanent modern-day societal interaction with people from other cultures and backgrounds. This is a situation that arose through large-scale historical movements of populations and colonization across geographical borders and regions (Ezeife, 2011).

There are many Indigenous populations around the world. Some examples include the Canadian Aboriginal people, the American Native Indians, the Maya people of "Mesoamerica," the Brazil's Indian populations, the Mapuche Indians of Chile, the Aboriginal people of Australia, the Maori of New Zealand, the Twa people in Congo, the Hadza of Tanzania and the Maasai of Kenya, the San and Khoe people of Angola, Botswana, Namibia, and other countries in the southern tip of Africa, and the Igbos, Yorubas, Hausa/Fulani of Nigeria on the West Coast of Africa. The United Nations estimates that there are about 370 million Indigenous people (roughly 6% of the total world population) living in over 70 countries worldwide (Wikipedia, 2018).

In Canada, the 2016 census indicates that nearly 1.7 million people are identified as Aboriginal, which accounts for a 4.9% of the total Canadian population and "a breathtaking 42.5 per cent increase since 2006, a growth rate more than four times that of their non-Indigenous counterparts" (Kirkup, 2017, p. 1). With regard to China, "ethnic minorities" (consisting of the non-Han Chinese population) make up 8.49% of the population of mainland China. Officially, 55 ethnic minority groups are recognized in the PRC (People's Republic of China) (Wikipedia, 2018).

2 Pre-contact Indigenous Education

The term "pre-contact" is used here to refer to the period before Indigenous communities were exposed to contact and subsequent interaction with people from other (different) races, cultures, and backgrounds such as colonizers or colonists, and permanent settlers. Pre-contact education in Indigenous cultures was holistic, environmentally sourced, and traditionally focused. Essentially, it reflected and drew heavily from the seven principles of Indigenous peoples' worldviews summarized by Simpson (2000) as cited by Hart (2010), thus:

> First, knowledge is holistic, cyclic, and dependent upon relationships and connections to living and non-living beings and entities. Second, there are many truths, and these truths are dependent upon individual experiences. Third, everything is alive. Fourth, all things are equal. Fifth, the land is sacred. Sixth, the relationship between people and the spiritual world is important. Seventh, human beings are least important in the world. (p. 3)

From this espoused worldview, it becomes obvious that Indigenous peoples' knowledge and education revolve around mutual interaction with one another and the natural world, which they regard as a common, shared territory. It is in recognition of this situation that Castellano (2000) identified the major features that characterize Indigenous knowledge as person-oriented, orally expressed, experientially or environmentally based, and holistically delivered or implemented.

3 Differences Between Indigenous and Western (Eurocentric) Educational Paradigms/Models

In their study that contrasted the Indigenous/Aboriginal educational paradigm/model to the Eurocentric model, Hughes and More (1997) highlighted the following distinguishing factors:

- Aboriginal model is holistic (integrated), and Western is analytical.
- Cooperation is focused upon and promoted in the Aboriginal model, while the Western model encourages competition.
- Aboriginal is essentially informal, whereas Western is intensely structured.
- Aboriginal education is apprenticeship-prone (observation and imitation), while Western leans heavily on verbal interaction and instruction—both oral and written.
- Aboriginal model dwells on real-life performance in the (natural) environment, while Western adheres to practice in contrived settings.
- Aboriginal education focuses on the mastery of context-specific skills, whereas Western "seeks to teach abstract content-free principles which can be applied in new previously inexperienced situations" (p. 15).

4 Conflicts that Arose in the Educational Realm or Arena During the Post-contact Era

These differences immediately led to huge conflicts between the Indigenous societies and the European settlers/colonizers who interacted with and have lived among people of Indigenous cultural backgrounds in the post-contact (modern) era/society, as summarized by O'Connor (2009), as follows:

> The incursion of Western society on Indigenous peoples brought about many cultural and psychological disruptions in the flow of life in traditional societies. Since the inception of modern education in the villages (reserves), the curricula, policies, textbooks, language of instruction and administration have been in conflict with the Native cultural systems. The modern public schools are not made to accommodate differences in Native worldviews, but to impose another culture that is Western. This has had a confusing effect on the Native students. Alienation and identity crisis of youth and their continual search for meaning are a condition of Native life today. New images of modernity collide with traditional symbols, values and beliefs. (p. 14)

The cultural disruption impinging on Indigenous students resulted in the alienation of these students from modern-day schools, a situation that gave rise to the widely acknowledged poor performance in, and often complete dropout from, schools. The lack of Indigenous and ethnic minority cultural/traditional knowledge and perspectives in school curricula have been identified as significant factors in the poor performance and high dropout rates among Indigenous and ethnic minority students, especially in the realm of mathematics/science education (Binda, 2001; Bourque, Bouchamma, & Larose, 2010; Cajete, 1994; Dorman & Ferguson, 2004; Ezeife, 2014a; Friesen & Ezeife, 2009).

The language of instruction (often the language principally spoken in the mainstream culture) has also been found to negatively affect the interest and performance of minority students in school courses in several multicultural and multi-ethnic countries including China (Nelson, 2005; Zhou, 2010). The same applies to a situation identified by some researchers whereby teachers who teach in culturally diverse classrooms may not have the requisite background cultural or local knowledge to meaningfully address the needs of minority students in the class, as pointed out by Rong (2006) who commented on the situation in China.

From the foregoing (post-contact education), it is obvious that the "melting pot" approach (as opposed to the "mixing pot") was adopted not only in the development of curriculum materials, but also in the implementation of the developed curriculum in schools. Drawing on the work of Okafor (1984), Ezeife (2017) summarized the "melting"/"mixing" pot attributes and highlighted the benefits and obvious pitfalls of each approach in multi-ethnic societies such as Canada and China where Indigenous/ethnic minority students constitute a significant part of the school population, thus:

"Melting" pot

- Minority cultures/issues are "melted" into a dominant culture.
- Molding process adopted—conformity.
- Form of acculturation.
- Minorities lose their identities.

And for "Mixing" pot

- Cultures are not melted away but mixed.
- No significant/substantive loss of cultural identity.
- Cultures become subcultures within a uniting umbrella.
- The educative process serves as an avenue for inculcating:

    1. Mutual understanding.
    2. Mutual respect.
    3. Mutual fellowship.
    4. Inter-ethnic harmony.
    5. Cross-cultural knowledge and awareness.

Looking closely at the enumerated attributes, benefits, and pitfalls, there is no doubt that mathematics education in all its ramifications—including the development and implementation of curriculum materials in schools in culturally diverse societies—should be undertaken with a "mixing pot" (as opposed to the "melting" pot) orientation. Over the years, this has been persistently advocated by Indigenous groups, several educators, authors, and researchers in both Canada and China (Assembly of Manitoba Chiefs, 1999; Berkowitz, 2001; Demmert, 2011; Ezeife, 2002; Gaskell, 2003; Jegede & Aikenhead, 1999; Matang, 2001;

Papp, 2016; Peng & Song, 2014; Sicat & David, 2011; Smith, 1994; among others). The underlying theme stressed by these advocate groups is the felt need to integrate the abundant cultural knowledge, traditions, and lived experiences (the schema) of Indigenous and ethnic minority learners into the mathematics and science education curricula used to teach them in schools.

In China, Peng and Song (2014) highlighted the scores of mathematics-related cultural practices of ethnic minorities in China, evident in their architecture, modes of dressing, their drawings and counting units, calendars, religious beliefs, and associated practices. Furthermore, the authors gave several examples of mathematics-based everyday life practices, equipment, and tools associated with and traditionally used by the Uygur and Tibetan Chinese ethnic minorities such as the geometrically designed stoves they use for cooking, their *Ketman* (a digging tool), their cave houses, and their *Kariz* (a unique irrigation system). These mathematics-related cultural practices and traditional activities, the authors argue, could be seamlessly integrated into and beneficially utilized in the mathematics teaching curriculum. This approach, it has been stressed, would make mathematics and science education not only relevant, but also meaningful to the students.

In the current state of affairs, these students are subjected to a classroom teaching experience where they simply acquire a lot of mathematics content which has little or no relevance to them because the content is completely bereft of the Indigenous/ethnic minority students' everyday lifeworld, their background environmental and traditional content knowledge, and their rich culture, as depicted in Fig. 1 (Ezeife, 2014b). This is the "missing link" between Indigenous traditional education and the Western system of education dwelt upon by Doige (2003). It is only when this missing link, gap, or corridor is effectively bridged that the sustained change advocated by Bishop, Berryman, Wearmouth, Peter, and Clapham (2012)—who investigated issues impacting on the education of ethnic minority students of Māori origin in New Zealand—can be fruitfully attained.

5 Next Steps and Directions for the Future

Contemporary research literature has an abundance of suggested strategies which, if adopted, would help inject and integrate the Indigenous and minority students' lifeworld culture into the culture of school

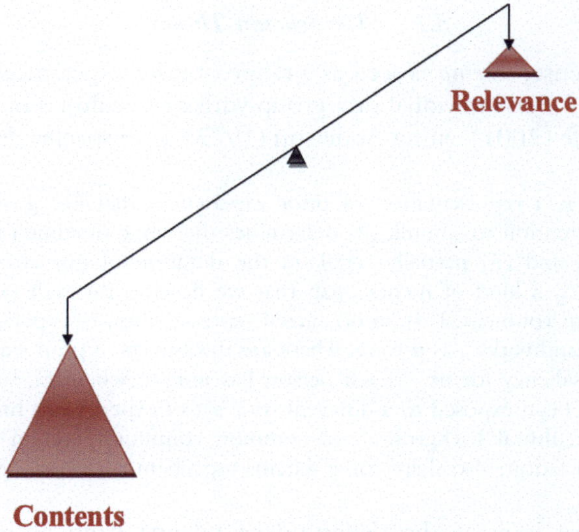

Fig. 1 Scale tilts heavily toward the "contents" arm, due to heavy "content," but little "relevance."

mathematics. For instance, there is a proposal for the utilization of the concepts of *cultural border crossing* and *collateral learning* in the teaching and learning of science/mathematics (Jegede & Aikenhead, 1999). Smooth border crossing—the ideal situation that would promote classroom learning maximally—is enabled (occurs) when students' life-world culture (derived from home, peer, societal, communal, traditional, environmental flora and fauna, etc.) is congruent with the culture of school mathematics. This enables the students (mathematics learners) to transition smoothly from their background lifeworld knowledge into the mathematics knowledge they are expected to acquire in school.

Thus, the knowledge the students bring from their lifeworld culture, for example, the background or culture-based knowledge they possess about a specific mathematical concept or a set of concepts, would facilitate the learning of a similar or related concept(s) in an actual classroom teaching/learning engagement, a situation described as simultaneous *collateral learning* (Ezeife, 2003).

5.1 *The Schema Theory*

In a broad sense, *schema* is seen as a range of prior experiences that have prominence for an individual or a group with close cultural or associative affinity. Ezeife (2001), citing Anderson (1972), gave specific details, thus:

> A schema is a representation of prior experiences that (1) governs our selective attention to stimuli, (2) determines the range of stimuli to which we attend, and (3) partially regulates the duration of our attention. A schema is ... a kind of mental map that we develop through experience with the environment. It is an organized representation of experience that contains "landmarks", as it were. These are the aspects of prior experiences that have saliency for us... Each person has unique schemata, since each person has been exposed to a different total set of experiences. But people of similar cultural background, of common community origin or small group association, also share some schema aspects in common. (p. 19)

An incisive look at the definition of schema from an educational perspective leads one to deduce that a person's schema is solidly bound to the person's cultural background, the environment, traditions, phenomena, etc. in which that individual grows up. Elaborating further, Ezeife (2001) stated:

> Children acquire the culture into which they are born, and their early learning in life is drawn directly from this cultural milieu. By the time the child is of school age, he or she has already formed impressions or conceptualizations of things around him or her, particularly natural events and phenomena. These internalized impressions are transferred to the school when the child eventually gets there. It is necessary, therefore, that the school continues along the path of learning that the youngster is treading, building on what the child has acquired from the home and environment. That way, the child sees school learning as culturally relevant, integrated with, and meaningful to the life ahead of him or her. Of course, there may be need, from time to time, for school learning to modify and re-direct the child's cultural or environmental learning, but definitely not to ignore, degrade, or dismiss it. (p. 21)

Based on research-sourced propositions, several of which have been effectively implemented in mathematics education programs targeted to Indigenous and ethnic minority learners (Ezeife, 2011, 2014a); this chapter takes the firm stance that mathematics education in culturally diverse countries like Canada and China should be solidly anchored on the integration of Indigenous mathematical knowledge, learning styles, traditional examples, and illustrations into extant school curricula, text books, and other mathematics teaching/learning resources. This approach, no doubt, will vivify mathematics teaching and learning for Indigenous and ethnic minority learners who would be emboldened to see mathematics as part of their lifeworld, and not as a foreign invention—a subject to avoid, and at the earliest opportunity, drop in school.

6 Some Actual Examples on the Utilization of Indigenous and Ethnic Minority Learners' Cultural Knowledge in Mathematics and Science Teaching/Learning Classrooms

6.1 The Use of the "Bow and Arrow" to Teach the "Conservation of Energy" Concept in Physics

To illustrate how students' schema can be advantageously tapped in the teaching of mathematical physics, let me narrate here my recent real-life experience. I was involved in a workshop, which included teaching some selected topics in mathematical physics to a group of Indigenous and ethnic minority students in one of my community projects in sub-Saharan Africa. One of the topics was *Conservation and Interconversion of Energy*. At some stage during the lesson, after the statement of the *Principle of Conservation of Energy*, and some simple experiments illustrating it, I was faced with the task of introducing and explaining the concept of *Interconversion of Energy*, that is, changing energy from one form to another. Initially, I was tempted to use a "Simple Cell" to illustrate how energy is converted from chemical to electrical form, and then to light and heat. This is the common illustration given in standard texts.

However, on second thought, I remembered that my class was not scientifically advanced, so to say, and might get confused and lose interest completely when the chemical equations involved in the process were presented. It was then that I decided to tap into the students' schema (with which I was familiar—having done some studies previously in the community), by choosing an illustration from their immediate environment. The illustration I chose was the "Bow and Arrow" (a commonly used sports and gaming equipment in this traditional community). With the students actively participating, we were able to establish that potential energy is stored in the bow when its string is fully stretched (that is, when strung out, extended, or drawn), and that this (potential energy) is converted into kinetic energy when the arrow is let go (shot) from the bow. Then, we discussed how on its way to a target (during flight) the kinetic energy of the arrow is used in doing work against the air viscosity, and finally on hitting a target the remainder of the kinetic energy is converted into sound (energy) and heat (energy)—which are clearly discernible by the observer because there is emission of sound (a thud is usually heard when the arrow strikes the target) and heat (the target struck by the arrow feels warm on touch).

The magical effect this illustration had on the class was manifested in the satisfied smiles and assenting nodes that greeted each stage of the teaching. One thought must have been going through the minds of the students in the class: "So, even our local 'bow and arrow' qualify as mathematical/scientific gadgets"? And what happened? During my next lesson the following day, the class size almost doubled. New participants had quickly enrolled in the program. The story of the bow and arrow (in a mathematics/science class) must have made its rounds! The conclusion to be drawn from this true-life experience is that if a direct appeal to (and apt tapping of) their schema can produce such a spontaneous, interest-invoking effect on a workshop group of Indigenous and ethnic minority students, then there is no limit to what can be accomplished by utilizing the same strategy when teaching a class of eager youngsters from the same cultural background in a typical mathematics and science class in the regular school setting. This is the future direction this chapter strongly suggests and advocates.

6.2 *The Use of Environmentally Sourced Materials and Illustrations in STEM Program/Classes*

There are several readily available natural resources (from the immediate environment of Indigenous and ethnic minority communities) that can be utilized in mathematics teaching/learning sessions, as exemplified by Ezeife's (2014a) effective use of a turtle shell and fern leaves (both of which were obtained from the Walpole Island Indigenous community in Ontario, Canada). These (and other resources shown in Fig. 3) were used in delivering mathematics and science lessons to Indigenous students enrolled in the catch-them-young 4-Winds STEM (Science Technology,

Engineering, and Mathematics) education program funded by the University of Windsor *Strategic Priority Fund* Project (2012–2014) and targeted at Aboriginal youth (Grades 5–8) in the Windsor-Essex County of Ontario, Canada (Ezeife, 2014a).

This weekend program was mounted over a three-year period at the Indigenous Education Centre (Turtle Island), University of Windsor, with the goal of attracting Indigenous and ethnic minority students to STEM fields—areas of study in which these students are known to be under-represented (Berkowitz, 2001; Binda, 2001; Friesen & Ezeife, 2009). The highly productive STEM program obviously achieved its goal because not only did it kindle the interest of the participants in STEM fields in the short run, but also the performance of most of the participating students considerably soared, leading to their career choices of mathematics and science-related disciplines in subsequent College and University pursuits—as authenticated by follow-up tracking records.

In Fig. 2, TV is a transversal cutting the two parallel lines shown with the double-edged arrows; the angles marked "a" (above the pair of parallel lines) are corresponding angles and are equal; similarly, the angles marked "b" (below the pair of parallel lines) are also corresponding angles and are equal. The sum of angles "a" and "b," (that is, $a + b) = 180^o$; so "a" and "b" are called supplementary angles (angles on the straight line, TV).

Fig. 2 Geometry—Transversals and Angles

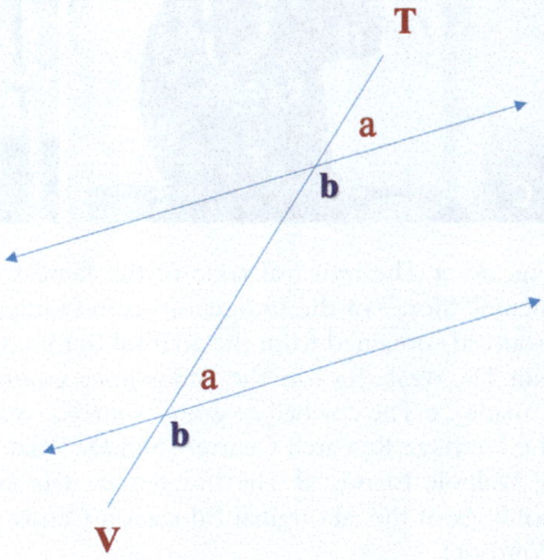

This Indigenous artifact (the turtle shell Fig. 3a) was effectively used to illustrate the same angular relationships in the concepts of transversals

Fig. 3 **a** The fern leaf (one of the familiar leaves identified with the environmental "flora" of the Indigenous community of Walpole Island), **b** A turtle shell (sourced/obtained from the archival holdings of the Heritage Research Centre—Nin Da Waab Jig—in the Indigenous community of Walpole Island, Ontario, Canada), **c** The cowbell or gong (sourced/obtained from the archival holdings of the Heritage Research Centre—Nin Da Waab Jig—in the Indigenous community of Walpole Island), **d** The Indigenous Tipi (sourced/obtained from the archival holdings of the Aboriginal Education Centre [Turtle Island] of the University of Windsor)

and parallel lines in Fig. 2 (as usually shown in typical mathematics text books) to a class of Indigenous students in the 4-Winds STEM project, University of Windsor. It was pointed out in class that the lines on both sides of the turtle shell could be seen to be parallel (almost), while the midrib of the shell is taken as the transversal.

Just like the turtle shell (Fig. 3a), the fern leaf (Fig. 3b)—a familiar sight to the students, since it is readily available in their environment/real life—was effectively used to illustrate the same angular relationships in the concepts of transversals and parallel lines to a class of Indigenous students in the 4-Winds STEM program, University of Windsor. It was pointed out in class that the lines on both sides of the leaf's midrib could be seen to be parallel, while the midrib is taken as the transversal. If the parallel lines meet the midrib at right angles, then the midrib is called a perpendicular transversal.

The cowbell (Fig. 3c), called *Godotaagan* in Ojibwe Indigenous language, is a familiar and commonly used artifact among Indigenous cultures. It was used to teach the concept of sound emission and transmission during the 4-Winds STEM program mounted for Aboriginal students at the University of Windsor.

The tipi (Fig. 3d), shown in part here, was assembled from scratch by the students in the 4-Winds STEM project under the guidance of the staff of the Aboriginal Education Centre, University of Windsor. It was then used to teach and illustrate several mathematics/science concepts during the 4-Winds STEM project such as shapes, areas, and volumes/capacity (Geometry); and temperature variations during the summer and winter months (Science).

In its discussion of the Tipi, the Wikipedia Encyclopedia (2018) states:

> Historically, the tipi has been used by Indigenous people of the Plains in the Great Plains and Canadian Prairies of North America ...The tipi is durable, provides warmth and comfort in winter, is cool in the heat of summer, and is dry during heavy rains. Tipis can be disassembled and packed away quickly when people need to relocate and can be reconstructed quickly upon settling in a new area. Historically, this portability was important to Plains Indians with their at-times nomadic lifestyle. (p. 1)

7 Summary, the Way Forward, and Conclusion

This chapter has drawn attention to, and discussed the research-authenticated low enrollment and poor performance in mathematics and science disciplines that affect at-risk learners—students of Indigenous and ethnic minority cultural backgrounds in several parts of the world, including Canada and China—the two countries focused upon in the chapter. Culturally oriented studies in contemporary research literature has drawn attention to the historical fact that Indigenous and ethnic minority people were students of nature, science, astronomy, and mathematics. Several authors have identified the wealth of experience and accomplishment of various Indigenous populations worldwide (Cajete 1994; Hatfield, Edwards, Bitter, & Morrow 2004; Smith 1994).

For instance, Smith (1994) narrated the case of the Skidi Pawnee—a highly reputed and widely known Indigenous group—who, in ancient times, were enthusiastic astronomers. They identified and described the planet Venus. This group "conceptualized the summer solstice" [and] "… in this way they could predict reoccurring solar phenomena" (p. 46). Similarly, Hatfield et al. (2004), in their work entitled *The giftedness of many cultures in mathematics*, highlighted the mathematical ingenuity of several Indigenous and ethnic minority cultures around the world. For example, in *Arithmetic*, specifically with regard to the system of numerals, the authors drew attention to the early use of a symbol for ten in place of 10 tally marks, and the invention of unit fractions by ancient Egyptians in Africa; the invention and introduction of a symbol for zero by Native Americans, and the related invention of negative numbers by ancient Chinese people.

In some other areas/strands of mathematics such as *Data Management and Probability*, the same authors documented how the Mayans were able to predict eclipses by the systematic analysis of astronomical tables which they had developed and used over a span of several centuries. In the field of *Algebraic Concepts*, the authors documented how ancient ethnic minority African cultures invented rectangular coordinates and utilized the concept in producing scale drawings and star clocks.

This chapter has addressed the key questions: If their ancestors were so engaged and leaders in the fields of mathematics, science, astronomy, and related disciplines, why do Indigenous and ethnic minority students shy away from these areas of study these days? These are the fields of learning in which their ancestors had excelled? Why do the few who enroll in mathematics, science, and associated fields perform poorly in examinations? In the effort to answer these topical questions, I took a research-supported stance that there is evidence of marginalization of Indigenous and ethnic minority learners of mathematics and science in schools through the use of curricula, pedagogy, teaching styles, and language of instruction—to name just a few of the identified handicapping, debilitating, and inhibiting factors often cited by Indigenous/ethnic minority students. The chapter also drew attention to the several areas of variance between the dominant and Indigenous/ethnic minority cultures as they affect school learning in multicultural countries like Canada and China and harped on the urgent need to address the missing link (Doige, 2003) manifested in the lack of relevance (especially in mathematics and science fields) felt by Indigenous/ethnic minority students in the current school system.

It is the firm stance of this chapter that the use of environmentally sourced mathematics and science teaching/learning content materials, examples and illustrations as promoted in the chapter, and the adoption of holistic teaching curricula, would go a long way in attracting Indigenous/ethnic minority learners to mathematics/science fields and also would, hopefully, shore up their school performance and consequent career representation in these fields so critical to firm technological development and advancement in today's world.

The felt need to adequately provide for, to accommodate the expressed cultural teaching/learning needs of, and thereby to sustain the mathematical growth, and meaningful development of students from these cultural backgrounds in schools in these two multicultural societies (Canada and China) is reinforced when cognizance is taken of the fact that the growth rate of the Indigenous population in Canada is more than four times that of their non-Indigenous counterparts (Kirkup, 2017). In addition, the ethnic minority population in China stands at a tangible 8.49% of the entire mainland Chinese population (Wikipedia, 2018). That should be the reasonable way forward, the needful path to tow or tread, and the sagacious plan to follow—in the realization of the noble and desirable goal of providing meaningful and sustainable mathematics education for the soaring population of Indigenous and ethnic minority students in Canada and China, and indeed, other multicultural societies and countries all over the world.

References

Assembly of Manitoba Chiefs. (1999). *Manitoba first nations schools needs assessment report*. Winnipeg: Author.

Berkowitz, P. (2001). Western science meets Mi'kmaq knowledge. *University Affairs, 42*(10), 16–20.

Binda, K. P. (2001). Aboriginal education in comparative and global perspectives: What has research and practice done for Aboriginal education in Canada? *Canadian and International Education, 30*(1), 1–16.

Bishop, R., Berryman, M., Wearmouth, J., Peter, M., & Clapham, S. (2012). *Te Kotahitanga: Maintaining, replicating, and sustaining change*. Ministry of Education, NZ: Education counts. Retrieve from https://www.educationcounts.govt.nz/publications/series/9977/te-kotahitanga.

Bourque, J., Bouchamma, Y., & Larose, F. (2010). Aboriginal students' achievement in science education: The effect of teaching methods. *The Alberta Journal of Educational Research, 56*(1), 57–71.

Cajete, G. A. (1994). *Look to the mountain: An ecology of indigenous education.* Skyland, NC: Kivaki Press.

Castellano, M. B. (2000). Updating Aboriginal traditions of knowledge. In G. J. S. Dei, B. L. Hall, & D. G. Rosenburg (Eds.), *Indigenous knowledges in global contexts*. Toronto: University of Toronto Press.

Demmert, W. G. (2011). What is culture-based education? Understanding pedagogy and curriculum. In J. Reyhner, W. S. Gilbert, & L. Lockard (Eds.), *Honoring our heritage: Culturally appropriate approaches to Indigenous education* (pp. 1–9). Flagstaff, AZ: Northern Arizona University.

De Wolf, G. D., Gregg, R. J., Harris, B. P., & Scargill, M. H. (1998). *Gage Canadian dictionary*. Vancouver: Gage Educational Publishing Company.

Doige, L. A. (2003). A missing link: Between traditional Aboriginal education and the western system of education. *Canadian Journal of Native Education, 27*(2), 144–160.

Dorman, J. P., & Ferguson, J. M. (2004). Associations between students' perceptions of mathematics classroom environment and self-handicapping in Australian and Canadian high schools. *McGill Journal of Education, 39*(1), 69–86.

Ezeife, A. N. (2001). Integrating the learner's schema, culture, and environment into the science classroom: Some cases involving traditional and Aboriginal societies. *Canadian and International Education, 30*(1), 17–44.

Ezeife, A. N. (2002). Mathematics and culture nexus: The Interactions of culture and mathematics in an Aboriginal classroom. *International Education Journal, 3*(3), 176–187.

Ezeife, A. N. (2003). The pervading influence of cultural border crossing and collateral learning on the learner of science and mathematics. *Canadian Journal of Native Education, 27*(2), 179–194.

Ezeife, A. N. (2011). A cultural and environmental spin to mathematics education: Research implementation experience in a Canadian Aboriginal community. *First Nations Perspectives, 4*(1), 2–39.

Ezeife, A. N. (2014a, April 30). *Teaching with an indigenous flavour: The implementation of a culture-oriented and environmentally sourced mathematics and science education program for Canadian Aboriginal youth.* Paper presented at the Aboriginal Research Forum (Shawane Dagosiwin) conference, Winnipeg, Manitoba, Canada.

Ezeife, A. N. (2014b, April 7–8). *Injecting nature and the learner's environment into contemporary mathematics education: The complementary roles of the teacher and the learner.* A plenary paper presented at the 1st annual International Reciprocal Learning conference, University of Windsor, 2014.

Ezeife, A. N. (2017, January–April). Education as a process of cultural "imbibition". *Lecture notes for the course, Aboriginal Ways of Knowing: Cultural, political and linguistic contexts.* University of Windsor.

Friesen, J. B., & Ezeife, A. N. (2009). Making science assessment culturally valid for Aboriginal students. *Canadian Journal of Native Education, 32*(2), 24–37.

Gaskell, J. (2003). Engaging science education within diverse cultures. *Curriculum Inquiry, 23*(3), 235–249.

Hart, M. A. (2010). Indigenous worldviews, knowledge, and research: The development of an Indigenous research paradigm. *Journal of Indigenous Voices in Social Work, 1*(1), 1–16.

Hatfield, M. M., Edwards, N. T., Bitter, G. G., & Morrow, J. (2004). *Mathematics methods for elementary and middle school teachers.* New York: Wiley.

Hitchcock, R. K., & Vinding, D. (2004). *Indigenous peoples' rights in southern Africa.* Retrieved from https://www.cambridge.org/core/journals/african-studies-review/article/robert-k-hitchcock-and-diana-vinding-eds-indigenous-peoples-rights-of-southern-africa-copenhagen-iwgia-2004-iwgia-doc-no-110-278.

Hughes, P., & More, A. J. (1997, December 4). *Aboriginal ways of learning and learning styles.* Paper presented at the annual conference of the Australian Association for Research in Education, Brisbane, Australia.

Jegede, O. J., & Aikenhead, G. (1999). Transcending cultural borders: Implications for science teaching. *Journal for Science and Technology Education, 17,* 45–66.

Kirkup, K. (2017, October 25). *Canada's indigenous population growing 4 times faster than rest of country.* Ottawa: The Canadian Press.

Matang, R. A. (2001, July 2–6). *The role of Ethnomathematics in mathematics education: Implications for mathematics curriculum in Papua New Guinea.*

Paper presented at the 11th World Congress of Comparative Education Societies (WCCES), Korea National University of Education, Chungbuk, South Korea.

Nelson, K. (2005). *Language policies and minority resistance in China*. Retrieved from https://s3.amazonaws.com/academia.edu.documents/38663080/volume_3.pdf?AWSAccessKeyId = AKIAIWOWYYGZ2Y53UL3A&Expires = 1538103463&Signature = H1FJZzMTt2l%2BUgL758VTWUS9wOo%3D&response-content-disposition = inline%3B%20filename%3DZakharia_Z._and_Arnstein_T._Eds_._2005_pdf#page = 35.

O'Connor, K. B. (2009). *Puzzles rather than answers: Co-construction a pedagogy of experiential, place-based and critical learning in Indigenous education*. PhD thesis, McGill University. Retrieved from https://www.collectionscanada.gc.ca/obj/thesescanada/vol1/QMM/TC-QMM-40708.pdf.

Okafor, F.C. (1984). Philosophy of Education and Third World Perspective. Enugu, Nigeria: Star Publishing Company.

Papp, T. (2016). Teacher strategies to improve education outcomes for indigenous students. *Comparative and International Education, 45*(3), 1–14.

Peng, A., & Song, N. (2014). The status quo development of mathematics education for ethnic minorities in China. *Studies in Ethnicity and Nationalism, 14*(1), 171–187.

Rong, M. (2006). *Education of ethnic minorities in contemporary China*. Retrieved from www.case.edu/affil/tibet/more.Tibetinfo/documents.

Sicat, L. V., & David, M. D. (2011). *Performance in basic mathematics of Indigenous students*. 2011 International Conference on Social Science and Humanity. IPEDR, 5. Retrieved from http://www.ipedr.com/vol5/no2/75-H10196.pdf.

Simpson, L. (2000). Anishinaabe ways of knowing. In J. Oakes, R. Riew, S. Koolage, L. Simpson, & N. Schuster (Eds.), Aboriginal health, identity and resources (pp. 165–185). Winnipeg, Manitoba, Canada: Native Studies Press.

Smith, M. R. (1994). Scientific knowledge and cultural knowledge in the classroom. In K. P. Binda (Ed.), *Critical issues in First Nations Education* (pp. 38–54). Brandon: BUNTEP, Faculty of Education, Brandon University, Brandon, Manitoba, Canada.

Wikipedia Encyclopedia. (2018, September 22). *Indigenous peoples*. Retrieved from https://en.wikipedia.org/wiki/Indigenous_peoples#United_Nations.

Wikipedia Encyclopedia. (2018, September 26). *Tipi*. Retrieved from https://en.wikipedia.org/wiki/Tipi.

Zhou, M. (2010). Language policy and illiteracy in ethnic minority communities in China. *Journal of Multilingual and Multicultural Development, 21*(2), 129–148. Retrieved from https://doi.org/10.1080/01434630008666398.

CHAPTER 15

A Case Study of the Change of Teacher's Teaching Belief Based on International Understanding

Wenping Zhu, Tebatso Namanyane, and Mengdie Li

1 Introduction

International understanding is the need to develop students' core literacy. Cultivating a nationality's international understanding is not only one of the important measures taken by the state and local governments to promote the internationalization of education, but also one of the important contents for cultivating students' core literacy and promoting the all-round developed people. It has new significance in current times and modern education.

International understanding of literacy is an important part of teachers' professional literacy. Cultivating and enhancing teachers' international

W. Zhu (✉) · T. Namanyane · M. Li
Southwest University, Chongqing, China
e-mail: 1987326742@qq.com

M. Li
e-mail: 1663652142@qq.com

S. C. Zhu et al. (eds.), *Reciprocal Learning for Cross-Cultural Mathematics Education*, Intercultural Reciprocal Learning in Chinese and Western Education,
https://doi.org/10.1007/978-3-030-56838-2_15

understanding literacy can optimize their international understanding behavior, increase the teachers' and students' international experience, and exercise their international communication abilities and skills. Therefore, international understanding of literacy can affect the professional development of teachers.

Literacy is holistic, and its meaning includes multiple levels of knowledge, ability, and attitude (Cai, 2011). Teachers' international understanding literacy is a comprehensive quality of international cognition, experience accumulation, and behavior improvement that teachers have displayed when they are participating in international understanding activities. It is a comprehensive quality that covers teachers' knowledge, emotion, and skill fields. Due to the rich content of teachers' teaching belief, our exploration of teachers' teaching belief examines a two-dimensional model for teacher belief, focusing on "what is teaching" and "how to teach."

When understanding the "what," we mainly consider two aspects: teaching objectives and teaching contents. The analysis of "how to teach" mainly explores the two aspects of teaching process and results. The understanding of teaching belief is mainly discussed from the perspectives of curriculum, student learning, and teacher teaching.

An international reciprocal learning project provides the conditions for this study (Xu & Connelly, 2013). The authors have participated in the international reciprocal learning research project between China and Canada. As a member of the project at Southwest University Mathematics Education Group, the authors have kept connections with teacher Z and have gotten to know her over the past two years. The authors have observed her performance when she is participating in the international reciprocal learning project between China and Canada, and have collected the data on her development of international beliefs, which have enhanced her understanding of international literacy gradually. This chapter describes three lessons taught by teacher Z. By observing, analyzing, and comparing teacher Z's teaching activities, this chapter summarizes her performances and experiences based on the international understanding.

2 Research Methods

The main research methods used in this study are: observations of classroom, interviews with the teacher, and video recordings of the classroom. As this is case study research, the analysis of the data will follow the case analysis process.

2.1 *Observation*

This study used classroom observations of teacher Z at R primary school in Chongqing, China, to understand the performance and influence of the teacher's teaching belief based on international understanding. Her classroom was observed five times over a period of three months.

2.2 *Interview*

This study used a semi-structured interview process to interview the mathematics teacher Z in Chongqing R Elementary School. Through her interviews, we learned about the basic situation of the schools, teachers, and students who participated in the China and Canada exchanges. We learned about her understanding of mathematics courses, teacher teaching, and student learning. At the same time, we collected her thoughts after three teaching activities and obtained her views on the implementation of international understanding in mathematics teaching in primary schools, so as to understand the characteristics and changes of her teaching belief.

2.3 *Data Analysis*

Teacher Z, who teaches at R Elementary School, is the main focus of investigation. Teacher Z is a female first-level mathematics teacher. She has a university bachelor degree. Teacher Z began to participate in the Chinese–Canadian reciprocal international cooperation project from October 2015. In this chapter, we analyze the teaching belief by analyzing the three stages of the teacher Z: before, one year into participation, and two years into participation. This chapter describes the changes of her teaching belief when she is participating in the international cooperation project. Through analyzing her three lessons, we can study the ways and impacts of his teaching belief transition.

2.4 *Video Analysis*

Based on the video coding and video case study methods of TIMSS Video Study, the teacher Z's classroom teaching video is analyzed, mainly from three aspects: curriculum, student learning, and teaching behavior, so as to analyze the changes of her teaching belief and find out the impacts of international understanding literacy on teaching belief.

3 Findings

In this section, we describe the three classes that teacher Z taught in mathematics. The first class focused on vertical and parallel lines. The second class focused on elapsed time and the third class focused on using estimation to solve problems.

3.1 *The Lesson on "Vertical and Parallel" Lines*

Teacher Z taught this lesson in March 2015. At that time, the mathematics teaching group of R primary school had not participated in the Chinese–Canadian Reciprocal Learning Project. Teacher R did not know much about the current teaching situation in foreign countries, especially the teaching of Canadian teachers. Her teachings at that time were mostly based on her local experiences. The following Table 1 summarizes the lesson.

3.1.1 *Discussion of R's Class*

When analyzing the characteristics of Teacher Z during lesson one using our analysis framework, we find the following:

What is the goal?

- Mathematics course: Have a clear goal (based on the implementation of knowledge points).
- Students' learning: Learning and mastering the knowledge of exams. Getting good grades is the driving force for learning. Using learning materials to solve problems means achieving learning goals.
- Teachers' teaching: Pay attention to the external teaching of knowledge and coping with the test (belonging to individual infusion) (Zheng, Fang, &Wang, 2015).

Table 1 Pre-international Canada–China reciprocal learning project participation: Lesson plan and analysis of lesson One: Vertical and Parallel

Link	*Teacher's activities*	*Students' activities*	*Analysis*
Intro	(Introduce): In this lesson, let's learn vertical and parallel. (She was drawing on the blackboard: Vertical and Parallel)	Listen to the teacher Z and look at the blackboard, then enter the learning state	It was believed that teachers can ignore the teaching introduction and can begin to the classroom teaching directly
Overall perception	(Question): Imagine two pencils are falling to the ground. What might happen? Draw what you think on the paper (Explanation): Both lines are drawn on the same paper. We can say they are in the same plane. Today, we will study the positional relationships of two straight lines in the same plane	Looking at the teacher and guessing what might happen. Listening to the teacher's explanations	It is believed that teachers should teach around the contents of books. Teaching should be based on the teacher's explanation. It was believed that students can spend less time exploring knowledge

(continued)

Table 1 (continued)

Link	*Teacher's activities*	*Students' activities*	*Analysis*
Cooperative inquiry In conclusion	1. Began to making a classification discussion to explore characteristics (Question): You can put your work together with your tablemate. Can you classify the two lines according to their positional characteristics? Why is this so? (Question): What will happen if you extend the two straight lines which you think they do not intersect? Bring your extended graphics and show them to us. What did you find? (Summary): In the same plane, intersecting, extending, and intersecting should be classified as intersecting, and the other is disjoint	report (Answer): Intersecting and dis intersecting	It is believed that the teacher is the main body in the curriculum implementation. The teacher leads the whole teaching process. The students' learning activities need the teacher to guide them It is believed that teaching should be carried out around the knowledge points. Students' mastery of knowledge points is the focus of learning

Link	*Teacher's activities*	*Students' activities*	*Analysis*
	2, Inductive understandings and clear the meanings of parallel and vertical (1) Reveal the concept of parallelism (Explanation): Firstly, let's study the situation of dis intersecting Look at this set of lines. Will they intersect? Please have a group of 4 people to discuss. (see student used three methods: 1, number of grids; 2, measurement; 3, extension method) (Question): How do you know that their distances are equal? Teachers and students use the methods of count, quantity and courseware to operate the above three methods again (Summary): Their positional relationship of two straight lines like this is called parallel. And these two straight lines are called parallel lines	(Answer): They will not intersect. Because the grids are same, the distances are equal (Answer): Use measurement; can also use extended method; will not intersect after extension	The understanding of teaching implementation: Thinking that the method can be based on the previous knowledge and experience. The second method of measuring requires the teacher to guide the students to complete, but it is mainly the teacher's hands-on operation, the students could not do it manually. The way to extend, students can draw conclusions by simply imagining and watching the teacher's courseware demonstration

(continued)

Table 1 (continued)

Link	*Teacher's activities*	*Students' activities*	*Analysis*
Introducing new knowledge Overall perception	(2) Let's study the situation where two straight lines intersect (Demo): The course was presented a set of intersecting lines that explain the meaning of the intersection. Show two lines that intersect at an acute angle and a right angle (Question): Which corners did the intersecting two lines form at the intersection? (Summary): Explain the definition of two lines perpendicular to each other (Question): Who can say how the two straight lines are perpendicular to each other? Let's take a look at how the book is describe (Question): Can you draw two straight lines perpendicular to each other by hand? (Explanation): Vertical is a special case of intersection. Now we use the knowledge we just learned to solve the problem	Follow the teacher to learn the meaning of the intersection (Answer): They can draw into an acute angle or right angle Reading the theorem on the book with the teacher (hands-on operation): Vertical alignment by hand Listen to the teacher's speaking	It is believed that the mathematics curriculum focuses on the explanation and guidance of the knowledge points of the textbooks. It is thought that the students' thinking is only about the teacher's questions She thought that the course is focused on helping students to master the knowledge. The evaluation of students' learning should be based on their understanding of the knowledge points on the textbooks
Practice and consolidation	(Exhibition): Please see Question 1 and Question 2 on P68 of this book (Expansion): Can you find vertical and parallel phenomena in your life? I also prepared some pictures. Can you find the vertical and parallel lines? Use your hands to draw it	Do the exercises after class (Exercise): Find the parallel and vertical lines in the picture and use their own hands to draw	It is believed that teacher should pay attention to the teaching of the theorems in the book. And the process of letting students make their own reasoning and verification can be neglected

Link	*Teacher's activities*	*Students' activities*	*Analysis*
Review and communication	(Summary of the whole lesson): As long as we are interested, we can find vertical and parallel in our lives (Question): What have you learned through today's study?	(Answer): Understand the meaning of vertical and parallel, and know the transitivity of parallel lines...	It is believed that the teaching summary should be extended and practiced around the knowledge gained

What is the content?

- Mathematics: Teaching materials and tasks are predesigned and contain few elements of international understanding.
- Students' learning: The knowledge that all learners have to learn is same. Learning is mainly about the understanding and mastery of certain, simple, and textual knowledge points.
- Teachers' teaching: The single, certain, preset textbook content and syllabus is the main content.

The process of teaching

- Mathematics course: In the implementation of the curriculum, teachers' activities play a decisive role. Teachers should have more activities than students.
- Students' learning: Students are accepting knowledge passively, and their activities are not important. The interaction between teachers and students in the learning process is of little significance. Teachers should pay attention to guide students to answer teachers' questions.
- Teachers' teaching: The teacher's control of the class is the key to teaching. Teaching is a one-way process from teaching to learning. Teachers should pay attention to the advancements of all aspects of the preset teaching. The teaching situation is not important.

The results of the lesson

- Mathematics course: Paying more attention to the teachers' teaching effects than the students' learning effects.
- Students' learning: The teaching results are mainly the students' learning and mastering of knowledge. Students' learning focuses on memorizing knowledge points or copying problem-solving methods.
- Teachers' teaching: Focus on the solution process of standard answers. Focus on oral evaluations and written test methods. Emphasis on the results of academic performance evaluation.

3.2 The Lesson on "Calculate Simple Elapsed Time"

This lesson is taught by Teacher Z in November 2016, when she had been involved in the China-Canada Reciprocal Learning Project for one year. She had gained contexts and understanding of Canadian teachers' teaching by participating in video conferences between China and Canada, going to Canadian elementary school classrooms, studying the curriculum standards of both two countries, and participating in the school's teaching seminars. Her teaching belief had transformed in that year. Her teaching is summarized in Table 2.

3.2.1 Discussion of R's Class

In the process of participating in the Chinese–Canadian Reciprocal International Cooperation Project, teacher Z gradually deepened her understanding of the Canadian and Chinese teachers' teaching. She also gradually changed her teaching belief. The teaching belief she held in this lesson is as follows:

What is the goal?

- Mathematics course: The mathematics curriculum forces on developing students' abilities.
- Students' learning: Students can enhance their international understanding through teacher's changes.
- Teachers' teaching: Teaching is not only to teach the student about the basic knowledge, basic skills, basic ideas, and basic activities that are necessary for their developments, but also to develop their mathematical thinking and learning qualities.

What is the content?

- Mathematics course: The content of the course gradually penetrates the elements of international understanding and advocates multi-integration.
- Advocate the generative content of the course.
- Students' learning: Learning content is adjustable and interrelated. Learning content can penetrate the theme of international understanding to enrich the student experience.
- Teachers' teaching: The teaching content should not be fixed knowledge, but open and varied. It should be updated with the

Table 2 One year into Canada–China reciprocal learning project: Lesson plan and analysis of lesson Two: Calculate Simple Elapsed Time

Link	*Teacher's activities*	*Students' activities*	*Analysis*
Classroom import	First, create a situation to introduce a new lesson (Explanation): Today is March 16th, 2017, and the current time is 9:10. Canadian teachers are taking classes with us today. Let's welcome them. Now, let's read the subject together!	(Read): Calculate simple elapsed time	The teaching content is drawn from life. She uses the Canadian teachers' coming to school as a case to introduce the teaching content. The content of the course is linked with the actual life, which can mobilize the students' interests in learning
Overall perception	Second, cooperation and exploration (Exhibition): What mathematical information can you get from the two pictures on the PPT? (Question): Can you solve this problem? Try to do it on the homework	(Answer): A Canadian teacher is going to school at 8:00 am on the left, and she is leaving school at 3:00 pm on the right. The question is how long does she stay at school?	Recognize that mathematics courses allow students to acquire knowledge through observation, conjecture, discussion, practice, and so on. Teachers' teaching gives students more space to think and explore

Link	*Teacher's activities*	*Students' activities*	*Analysis*
Cooperative inquiry	(Guiding): Communicate your methods with others in the group, and talk about your methods and reasons (Question): Who can share your methods and reasons with us?	(Discussion): Communicating with each other (Answer): Method 1: Use the method of calculating: 3 + 4 = 7 hours Method 2: Use the method of drawing line segments	The teaching belief has achieved two transformations: 1. The transformation of teaching design. She let students understand the principles of mathematics and know where mathematics comes from. 2. The transformation of teaching operations. She gives space to the students to let them have more time and space to think about different methods
Expand teaching	Listen to the students' answers and guide them to summarize the steps and principles of the different methods	(Discussion): Ask questions about the time (Answer): Method 3: Use the method of 24-hour time to calculate. 15–8 = 7 hours	Students are considered to be the discoverers and explorers of learning. They have the abilities to summarize the problems solutions through group discussions, guessing hypotheses, trying exercises and verifying reasoning

(continued)

Table 2 (continued)

Link	*Teacher's activities*	*Students' activities*	*Analysis*
Interactions between teachers and students Cooperative inquiry	(Guiding): Raise the point which you don't understand and ask your classmates to help you answer it (Question): Do you have different practices? (Question): Do other students have any thoughts?	(Discussion): All students ask questions and solve problems with each other. Discuss how to use 24-hour calculating to represent time and time moment. They also discuss how to calculate it (Drilling): Communicating with each other about the difference between time and time moment. Summarizing the points which should be paid attention to in the 24-hour calculating method (Answer): Can also count with our fingers (show it to the whole class). Method 4: counting with fingers	Thought that students can use their own ideas to summarize and refine the solution The teachers' teaching task is to guide them to explore things under the teacher's instructional design It is believed that under the guidance of teachers, students can summarize Method 3 and Method 4 through discussion and practice. In this session, teachers should let the students explore knowledge by themselves
Summary Sublimation	(Explanation): These methods are the same whether using the segmented timing method or the 24-hour calculating method. Where is the "same" performance? (Explanation): Can you use the arriving time in the morning minus the leaving time in the afternoon? Why?	(Exploration): Think about the similarities between these methods. Both use the ending time minus the starting time is the elapsed time Explore the problem	It is believed that the mathematics curriculum should not only focus on the guidance of knowledge, but also pay attention to the enlightenment of students' thinking and methods. Teaching should focus on letting students understand and explore knowledge in their own area

Link	*Teacher's activities*	*Students' activities*	*Analysis*
Classroom exercise In conclusion	(Explanation): Let's solve the problems with these strategies we just learned (Exhibition) Question 1: The daily business hours of the hot pot restaurant Question 2: Children's sleeping time on the day of Sleeping Day on the World in different regions. Question 3: How long of this lesson (Question): What should you pay attention to when calculating the elapsed time? (Summary): What are your gains in this lesson?	(Exercise): Do exercises and share their own practices with classmates (Answer): Paying attention to the addition and subtraction of time values, the distinction between time and time moment, and the labeling of units (Review): Talk about their learning gains	Linking course learning to solving practical problems, focusing on testing students' abilities to apply what they have learned to solve practical problems It is believed that teachers can infiltrate some international knowledge into teaching, thus enriching students' knowledge

understanding of teachers and students and the changes in the situation.

The process of teaching

- Mathematics course: Course implementation should focus on giving students more space to explore.
- Students' learning: Learning is based on students' existing knowledge. Students have the abilities to use learning strategies to solve problems. Teachers should pay attention to training students' learning methods and cultivating their awareness of questions.
- Teachers' teaching: Teaching is regarded as an active process, and it can be adjusted to student's learning state in a timely manner. Teachers should encourage students to use creative learning and problem-solving methods. Teachers should pay attention to students' understanding of knowledge and problem solving.

The results of the lesson

- Mathematics course: Mathematical courses focus on the common developments of teachers and students.
- Students' learning: The knowledge acquired by learners is unique and diverse. Students have the abilities to acquire knowledge in their own way. Students' internal motivation and interests in learning are the driving force for learning.
- Teachers' teaching: Focus on the diverse answers of students. Focus on evaluating students' learning attitudes and learning methods. Emphasize the abilities of students to acquire and develop their personalities during the learning process. Teachers can test the students' learning outcomes in a variety of ways after they have finished their studies.

3.3 The Lesson of "Using Estimation to Solve Problems"

Teacher Z taught this lesson in October 2017. Since joining the China–Canada Reciprocal Learning Project in 2015, she has communicated with many teachers and students in Canada. In January 2017, she and the members of mathematics team of R Elementary School went to T

Elementary School in Canada to engage in professional development. Combined with her observations and reflections on the teaching of Canadian teachers, she has innovated her own teaching. The summary of the teaching is found in Table 3.

3.3.1 Discussion of the Third Class

The teaching belief that Z teachers held in this lesson has the following characteristics:

What is the goal?

- Mathematics course: The mathematics curriculum should cultivate students' mathematics abilities, thinking abilities, and comprehensive qualities.
- Students' learning: The development of students is a long-term process. Teachers should guide the long-term developments of students.
- Teachers' teaching: Teachers should create specific problem situations, so that students can really feel the connection between mathematics knowledge and real life.

What is the content?

- Mathematics course: The mathematics curriculum is linked to other courses. Teachers should focus on the integrations of the course contents.
- Students' learning: Students should be taught to develop the knowledge and skills, which they needed. They should be guided to understand the basics of mathematical knowledge.
- Teachers' teaching: Mathematical teaching is not only to teach students knowledge, but also to be extended to students' learning, life, and entertainment.

The process of teaching

- Mathematics course: Students are the masters of mathematics courses. Teachers are the guiders and designers of this curriculum.

Table 3 Two years into Canada–China reciprocal learning project participation: Lesson plan and analysis of lesson Three: Using estimation to Solve problems

Link	*Teacher's Activities*	*Students' Activities*	*Analysis*
Import	Please take out the analysis form and complete the first step	Students find information and problems, circle the numbers, outline and focus on keyword phrases, etc. according to their own understanding	Apply the teaching method of the Canadian side to the classroom teaching in China. She lets the students solve the problem independently
Understand the meaning of the question	(Question): What information do you know from the picture?	Answer the question Student 1: I know that a total of 182 pineapples have been picked. If 8 per box, the question is that are 18 cartons enough?	Using analysis sheets and paying attention to cultivating students' self-reflective abilities and cooperative inquiry consciousness in teaching
Guiding activity	(Explanation): We first use the estimation method to solve this problem. Please complete the second and third steps of the analysis Through using analysis sheets, students are led to find information and problems, understand the meaning of numbers and keyword phrases on their own	(Activities): (1) Complete the analysis forms independently (2) Exchange their analysis forms within the group (3) Present their analysis forms to the whole classmates	Teachers should respect the different answers of students. They should be good at listening to students' answers, and giving them sufficient time to think and discuss

Link	*Teacher's Activities*	*Students' Activities*	*Analysis*
Exchange discussion Option 1: by estimating litter comparison	(Question): Why do you estimate 182 to be smaller but not bigger? (Summary): When we are estimating, it is necessary to choose whether to estimate smaller or bigger according to the specific situation. It is also need to consider the estimated numbers for us to calculate	(Answer): $182 \div 8 \approx 20$ (a), we need 20 cartons even underestimated the number of pineapples, so 18 cartons are certainly not enough (Answer): If you estimate 182 as 160, you need 20 boxes, so 18 boxes are definitely not enough. Thinking 182 as 180, and the quotient is definitely greater than 20	Recognizing that students' learning is a proactive process. They can solve problems by thinking, exploring, cooperating, operating, computing, and guessing on their own
Option 2: by Estimating bigger comparison	(Question): What is he asking for and comparing? Why not treat 18 as a smaller that 15? (Summary): It is that we can only estimate the bigger one to compare in this question	(Answer): By comparing the total number of pineapples. There are 18 cartons, if every one for 8 pineapples, how many pineapples can you wear? $18 \times 8 \approx 160$ (pieces), you can only install 160 pineapples. So 182 pineapples will definitely not fit	Realize that students are able to choose whether to estimate larger or smaller comparison methods depending on the context. They can use the right numbers for easier computing on their own

(continued)

Table 3 (continued)

Link	*Teacher's Activities*	*Students' Activities*	*Analysis*
Option 3: Comparing each number	(Question): Is there a different way to solve this problem? (Question): What is he asking for and comparing with?	(Answer): If there are 182 pineapples will be packed in 18 cartons, it should be considered how many pineapples are placed in each carton? $182 \div 18 \approx 10$ (pieces). If we think 182 as 180, each carton should hold 10 pineapples. While each box can only hold 8, 182 pineapples can certainly not be enough (Answer): Compare each number	In teaching, teachers can talk and do less. They should let students explore and understand by themselves
Summary review	(Question): Compare these three methods, what are the similarities and differences? (Explanation): The amount of comparison is different, but all draw the same conclusion that 18 cartons are not enough	(Answer): Both use the estimation method that find another quantity based on two quantities, and then compare it (Answer): The amount of comparison is different, and the strategy for solving problems is different	Realize that teaching should give students full opportunities to think and speak. Teacher should let every student participate in the class and express their own opinions
Option 4: Multiplication calculation	(Question): It seems that our estimate method is correct. Is there another different way?	(Answer): By comparing the total number of pineapples. Directly calculated: $18 \times 8 = 144$ (pieces), $144 < 162$, so 18 cartons are not enough	She believed that students have the abilities to solve problems through independent thinking and inquiry
Option 4: Division calculation	(Question): Are there any different ways?	(Answer): $182 \div 8 = 22$ pieces and have another 6 pieces , $22 + 1 = 22$ (pieces). Because $22 > 20$, it takes 22 cartons to be fully loaded, it is not enough Compare how many pineapples are packed in each carton (each serving number)	Students should be exposed to a complete inquiry process

Link	*Teacher's Activities*	*Students' Activities*	*Analysis*
Summary communication	(Question): What do you find by comparing these three different estimation methods? (Explanation): When we solve the problem, whether use the estimate or calculate method should depend on the actual situation. We should also use the estimate bigger or smaller method according to the specific problem	(Answer): The first comparison is the number of pineapples (total), the second comparison is the number of cartons (number of copies), and the third comparison is how many pineapples are required for each carton (each serving number)	Recognizing that teachers are participants in teaching. They should learn to listen, and they should be good at capturing and using every message, every kind of difference and every conflict

- Students' learning: Letting students experience the process of exploring and discovering knowledge will be more conducive to their lifelong development.
- Teachers' teaching: Teaching should focus on guiding students to build new knowledge independently. Teachers should pay attention to the long-term cultivation of students' abilities. And they should pay attention to the students' development needs at different stages.

The results of the lesson

- Mathematics course: The mathematics curriculum is not only about the transfer of knowledge points, but also to teach students the basic ideas, skills, and students' core literacy.
- Students' learning: Students should be inspired and guided to eventually become the masters of studying by the teachers.
- Teachers' teaching: Through the transformation of teachers' teaching belief and behaviors, both teachers and students will gain the necessary qualities and key abilities.

4 Discussion and Conclusions

International understanding of literacy has a major impact on teachers' teaching belief. The transformation of teachers' teaching belief requires certain changes of conditions, processes, and levels. The transformation of teachers' teaching belief mainly includes their understandings of four aspects, which are teaching objectives, content, process, and results. In order to optimize the teachers' teaching belief to improve their teaching practice levels, the suggestions put forward in this study are: to expand the teachers' teaching horizon; to strengthen international understanding in international exchanges; to promote the transformation of teachers' teaching belief; and to improve the levels of international understanding literacy.

4.1 The Changes and Reasons of Teacher Z's Teaching Belief

It is believed that teacher Z's teaching belief has changed through participating in the Chinese–Canadian Reciprocal Learning Project. The following are the key changes that occurred in teacher Z's thinking when considering aspects such as: what the goal of teaching is? what the content is? the process of teaching, and the results of the lesson.

What is the goal?

- Mathematics course: From focusing on the implementation of knowledge points to focusing on the cultivation of students' abilities and literacies.
- Students' learning: From focusing on the improvements of students' mathematics performance to focusing on the cultivation of students' essential qualities and key abilities.
- Teachers' teaching: From the main response to the examination to the development of students' mathematical abilities.

What is the content?

- Mathematics course: From focusing on presupposition contents to focusing on generative contents.
- Students' learning: From focusing on textual contents to focusing on principled contents.
- Teachers' teaching: From fixed and preset outline contents to comprehensive and generative contents.

The process of teaching

- Mathematics course: From the teacher's leadership in the classroom to the students' leadership of learning in the classroom.
- Students' learning: From paying attention to students' mastery of knowledge to focusing on the cultivations of students' abilities.
- Teachers' teaching: From focusing on the teacher's teaching process to focusing on the students' learning process.

The results of the lesson

- Mathematics course: From focusing on the implementation of the knowledge points to focusing on the comprehensive qualities of teachers and students and the cultivation of their core literacies.

- Students' learning: From focusing on students' memory and practice to focusing on the developments of their internal motivations and interests in learning.
- Teachers' teaching: From focusing on teachers' teaching effects to focusing on the students' improvements about learning qualities and abilities.

By analyzing the changes of teacher Z's teaching belief in the teaching objectives, content, process, and results, we believe that the reasons for these changes can be summarized into two categories. On the one hand, teachers can form thinking innovation in the process of expanding the horizon when they are participating in the China and Canada Reciprocal Learning Project. For these three years, teacher Z has taken part in discussing, exploring, and thinking about this project work. She pondered the differences and connections between the Chinese and Canadian sides in terms of teaching objectives, process, and belief. She had discussed the topics of mutual interests with Canadian teachers and deepened her understandings of the teaching that occurs in these two countries.

Combined with her research interests, she went to the Canadian classrooms for field study and in-depth observation. She has collected, organized, and analyzed the materials. She often applies new insights which she has gained from the project to her practices and further enhances her teaching skills. The project platform has provided supports for her to broaden her vision and innovative thinking.

On the other hand, teachers can compare the similarities and differences between the Chinese and Canadian teaching strategies to improve the teaching practices on the basis of understanding these principles. Teacher Z often studies and thinks when participating in the international cooperation projects. On the basis of exploring "what it is," she further thinks about "why it is," which includes question such as "why does the teacher teach in this way? what are the mathematical principles behind it? why are the students using this method to solve the problems? what is their knowledge base and thinking characteristics? are there differences between teachers and students in China and Canada in terms of their level of thinking and knowledge base, and what are the reasons for this phenomenon?" (Teacher Z). Teacher Z has learned about the teaching principles of both China and Canada in this process of exploration from "what is" to "why." She has enriched her teaching knowledge, improved her ideas and thinking levels, and cultivated abilities in this process.

4.2 *The Paths to Enhance Teachers' International Literacy Understandings and to Promote Their Teaching Belief Transformations*

We should pay attention to the transformations of teachers' teaching belief and ways to cultivate their international understanding. The first way is to expand the teachers' horizon. The teaching horizon can influence a teacher's thinking and practice level. We can use the following three strategies to expand the teacher's teaching horizon. The first strategy is to recognize the values of international understanding literacy. International understanding literacy, as a necessary core literacy for human development in the twenty-first century, deserves everyone's attention and to be put into practice. Teachers should connect it with teaching activities in light of specific situations, so as to achieve the renewals of teaching belief and the improvements of behaviors.

The second strategy is to reflect on the educational practices of different cultures. In the face of teachings in different countries, teachers need to reflect on the rationality and feasibility of this educational practice in light of their cultural background. Their process of contrasting, reflecting, and learning from different cultures is also the process in which they further enhance international understanding.

The third strategy is to train teachers to understand literacy internationally. On the level of understanding, we must pay attention to the meaning and value of international understanding. On the level of practice, we should combine it with teaching practices and conditions. We should form new teaching belief through practice and reflection to serve those teaching practices.

The second way is to strengthen the international understanding in the progress of international exchanges. The change of teachers' teaching belief must go through three stages: identifying differences, understanding similarities and differences, and learning from experiences. First of all, we should identify differences and be inclusive. Teachers should treat these differences correctly when participating in internationally understood activities. Not only should they pay attention to what the differences are, but also the causes and implications of these differences. It is necessary to think about how to integrate the belief, ideas, methods, and experiences that they learn, so as to prepare for further practice.

The second stage is to understand the similarities and differences and then learn from each other. The generation and occurrence of different beliefs need to be attached to certain situations. While respecting the differences, teachers should also understand the situations created by these differences. They should carry out certain imitations, practices, and integrations in light of specific realities. The third stage is to learn from experiences and to introduce new ideas. The transformation of teachers' teaching belief requires them to be based on teaching practices, and through reflection, summarization and excellent experiences, to improve the levels of teaching practice.

The third way is to promote the transformation of teachers' teaching belief. First, we need to create conditions to transform teachers' teaching belief. The transformation of teachers' teaching belief requires certain internal conditions and external conditions. The internal conditions are mainly obtained through the teachers' own efforts, which include possessing certain educational and teaching knowledge, educational psychology knowledge, teaching method knowledge. Furthermore, accumulating rich teaching experiences enhances teachers' own teaching design and reflective skills. The external conditions mainly depend on the teaching environment in which the teacher is located, for example, having a place and medium that enhances international understanding and implementation of international understanding activities; a research team capable of providing teaching support and scientific research; and the material and financial support which are necessary to change the belief of teaching.

Following the transformation of beliefs, teachers will encounter various specific problems. This requires teachers to base themselves on realities and combine teachings with practice to lay the foundation for the development of teachers and students. Third, combine teaching reflection and practice to optimize teaching. Teachers need to work hard to enrich their professional knowledge reserves, strengthen teaching reflection, and thus prepare for the improvements and optimizations of teaching belief.

5 Limitations

This study mainly analyzes the changes of teaching belief from three specific examples of teaching, so there may be biases in understanding. Limited by the research conditions and levels, this chapter only focuses on the teacher Z, a teacher who participates in the China and Canada Reciprocal Learning Project, so the range of the research is limited.

The research conclusions may need further discussion and scrutiny. Therefore, the follow-up research needs to further study the teachers' teaching belief in other disciplines and other grades in order to draw more reasonable conclusions and suggestions.

References

Cai, Q. (2011). *Literacy: DNA of curriculum reform* (p. 42). Taipei City: Higher Education Culture Enterprise Co.

Xu, S., & Connelly, F. M. (Project Directors). (2013). *Reciprocal learning in teacher education and school education between Canada and China.* Social Sciences and Humanities Research Council of Canada (SSHRC) 2013–2020 [Grant 895-2012-1011].

Zheng, J., Fang, J., & Wang, Z. (2015). Analysis of the influence of junior middle school teachers' teaching belief on their classroom management methods. *Teaching and Management, 1,* 38.

CHAPTER 16

Mathematics as a Cultural Role Player in School Development: Perspectives from the East and West

Anthony N. Ezeife

1 Background and Introduction

School development involves a wide-ranging series of appropriately planned activities and the coordinated efforts of several practitioners, functionaries, and stakeholders in the educational arena. Thus, adequate school development would entail the active involvement of school administrators, teachers, support staff, parents, guardians, and community members in creating a school climate and "environment that nurtures both students and adults" (Center for Effective Collaboration and Practice [CECP], 2001, p. 1). Essentially, a worthwhile or adequate school development program helps "schools focus their operations around effective child development and successful teaching and learning" putting "the

A. N. Ezeife (✉)
University of Windsor, Windsor, ON, Canada
e-mail: aezeife@uwindsor.ca

S. C. Zhu et al. (eds.), *Reciprocal Learning for Cross-Cultural Mathematics Education*, Intercultural Reciprocal Learning in Chinese and Western Education,
https://doi.org/10.1007/978-3-030-56838-2_16

development of ***all*** [emphasis in the original source] children at the centre of the educational process" (CECP, 2001, p. 1).

The emphasis on "all" children evokes the concept of inclusivity whereby learners from all cultural backgrounds are adequately provided for, and afforded an opportunity and learning climate, which would enable them to benefit maximally in school. Inclusivity, in turn, would entail developing and implementing a culture-sensitive curriculum that would address the specific, perceived, and prevalent needs of learners from non-mainstream cultural backgrounds in any school setting, thereby making every student feel valued and represented. This chapter will specifically address the issues of learners who are academically at risk in the subject area of mathematics with focus on Chinese ethnic minority mathematics students and their counterparts in Canada—mathematics learners of Canadian Indigenous cultural backgrounds.

2 Mathematics in Today's World

The importance of mathematics in today's world is widely acknowledged. Several authors, researchers, and research institutes have articulated the need and importance of robust mathematics education in contemporary society. For instance, the International Association for the Evaluation of Educational Achievement (IEA) stated pointedly in its 2012 report, which incorporated the 2011 Third International Mathematics and Science Study (TIMSS) results in mathematics, that:

> The world is becoming increasingly quantified and all students need to be well grounded in mathematical and technological thinking to live a productive life. To be effective future citizens, students need mathematics to understand daily news and grasp world events, often described through statistics, increases, and decreases [...] Mathematics is the foundation for further study in a number of school subjects, most notably the sciences; and mathematics problem solving builds logical reasoning skills that can be applied in many situations. (Arora, Foy, Martin, & Mullis, 2012, p. 25)

Still dwelling on the importance of mathematics, the National Council of Teachers of Mathematics (NCTM) made the following emphatic statement: "In this changing world, those who understand and can do mathematics will have significantly enhanced opportunities and options for shaping their futures. A lack of mathematical competence keeps these doors closed" (NCTM, 2000, p. 50).

Despite its well-articulated and widely acknowledged importance, a good number of students still shy away from mathematics, remain aloof to the need for it, perform poorly in it, and drop it at the earliest opportunity; they often pay a bitter price for this decision in their future careers. This mathematics phobia, with its attendant malaise—low enrolment in mathematics and related fields of study—has remained the unflattering state of affairs in high school education in many parts of the world for several decades, especially among students from minority and Indigenous cultural backgrounds (Binda, 2001; Gaskell, 2003; Snyder & Dillow, 2011). In the 1980s, the National Research Council (NRC) succinctly described the status quo when it lamented in its report: "Mathematics is the worst curricular villain in driving students to failure in school. When math acts as a filter, it not only filters students out of careers, but frequently out of school itself" (NRC, 1989, p. 7). This situation, so aptly described by the NRC in the 1980s, has stubbornly persisted and is even worsening, as confirmed by recent research findings in different regions of the world where authors and researchers report about the relatively low enrolment, difficulties experienced by learners, poor performance in examinations, and high dropout rates of high school students in mathematics, science, and other technologically oriented fields (Bourque, Bouchamma, & Larose, 2010; Friesen & Ezeife, 2009; Lauangrath & Vilaythong, 2010; among others).

Many of the authors/researchers have, therefore, called for a reorientation in school development programs—with specific attention to mathematics curriculum content, and a redirection in classroom implementation procedures and practices. The advocated reorientation is intended to make mathematics teaching and learning more meaningful, effective, and relevant to learners, especially at-risk learners from ethnic minority and Indigenous cultural backgrounds—so as to positively realign their disposition and attitude to mathematics education, with the overall goal of attaining a measure of parity with regard to their enrolment and performance in mathematics and related disciplines—with their counterparts from mainstream cultural backgrounds (Moreno-Garcia, 2012; Thomson, 2009). The relatively poor performance of many Indigenous students in mathematics has been a running trend dating back several years, as authenticated by research. For example, Binda (2001) revealed the following statistics from his study: In 1997, the mean mathematics performance of Manitoba Aboriginal students in First Nations' schools was 19.6% while the provincial average was 55.6%. In 1998, the mean score of Aboriginal students fell to 14.4% while the provincial average rose to 61.2%.

3 Why Do Many Students from Ethnic Minority and Indigenous Cultural Backgrounds Shy Away from and Perform Poorly in Mathematics?

One of the fundamental reasons given for the high dropout rate and poor mathematics performance of these at-risk learners is the lack of relevance of school mathematics content to the students' real-life experiences. In other words, the mathematics taught in school is bereft of ethnic minority and Indigenous cultural and environmental content (Sicat & David, 2011). Some other constraints or barriers to providing adequate mathematics education for ethnic minority and Indigenous students include the remoteness of several ethnic minority and Indigenous communities and settlements, inadequate teaching texts and related instructional materials, the language of instruction, and the way and manner mathematics is taught to these students, often by teachers who do not share the same cultural origins and values with the ethnic minority/Indigenous students they teach.

This is because there is usually a stark shortage of qualified minority teachers. Additionally, mainstream teachers' lack of local knowledge has also been identified as one of the obstacles to the meaningful education of minority students (Rong, 2006). Based on the findings of their study, Warren, Baturo, and Cooper (2004) commented on the issue of non-Indigenous teachers, especially those teaching Indigenous/minority students in remote rural areas:

> These non-Indigenous teachers and teacher-aides were not familiar with educational contexts in which Indigenous students learn and hence they tended to adopt traditional contextualised situations such as money, consumption, and measuring outside the classroom, and in many instances did not even reflect the remote and rural environments in which they were working. (p. 159)

4 Addressing Setbacks and Tackling Obvious Constraints and Barriers

To address the inadequacies and pitfalls, and thereby improve the status quo in the realm of ethnic minority/Indigenous education in general, and minority mathematics education in particular, several scholars, researchers, mathematics educators, and other practitioners in the field have stressed the need to integrate the learners' culture, environment, familiar everyday activities, and traditional practices into the mathematics curriculum. Following Ascher's (1998) ideas in her work on "Ethnomathematics," this chapter postulates that, for mathematics education to be really meaningful, effective, and relevant to students from ethnic minority and Indigenous cultural groups, the traditions of these learners and the ways they ascribe meaning to their environment and the social world in which they function must be fully reflected in the mathematics they are taught (and are expected to learn) in schools. Ascher stressed the importance of this approach to the teaching and learning of mathematics and affirmed that "mathematical ideas are cultural expressions."

This is the current and ever-growing discourse in the field of mathematics education, especially in countries and societies with multicultural populations, which include people of ethnic minority and Indigenous cultural origins. In their recent contribution to this discourse, Peng and Song (2014) stated,

> Given widespread concerns about equity in mathematics education, educators in countries with diverse multicultural populations have called for the recognition that mathematics is a cultural product, and, thus, mathematics education must take into account the growing diversity of students. (p. 172)

Both Canada and China—the two countries involved in the ongoing Reciprocal Learning Research Partnership Project—have diverse multicultural populations, of which ethnic minority and Indigenous peoples constitute a significant proportion. How can meaningful and effective mathematics education be incorporated into school development programs in these two countries under the auspices of the Reciprocal Learning Project? Essentially, what can researchers in the Mathematics Research Team of the Reciprocal Learning Project—in both China and

Canada—contribute toward the firm injection of the cultural mathematical practices and knowledge systems of ethnic minority and Indigenous students into Canada's and China's mathematics education programs? How can this integration goal be attained? These questions and the effort to seek answers to them fall squarely into one of the set goals—cultural perspectives—of the Mathematics Research Team. This chapter endeavors to seek answers to these important questions, providing supportive examples, illustrations, and projections for follow-up work, as appropriate.

5 What Can Be Done? How Can It Be Done?

Contemporary mathematics education research embarked upon by researchers in both the East–West educational systems has yielded a wealth of information with regard to the mathematics-oriented practices of ethnic minority/Indigenous mathematics learners. For example, in their recent work, Peng and Song (2014) stated, "There are various forms of mathematics underlying cultural practices of ethnic minorities in China, evident in architecture, dress, drawings, counting units, chronometers, methods of calendar calculation, and religious beliefs" (p. 176). The authors proceeded to give several examples of mathematics-related equipment and tools associated with, and traditionally used by the Uygur and Tibetan Chinese ethnic minorities:

> …distinct forms of mathematics have…been discovered in the everyday life practices of Uygur people. For example, the stove for cooking food constitutes a typical abutment structure; geometrical designs can be found in everyday tools such as the Ketman (a tool for digging) the Kariz (an irrigation system) and typical Uyghur (sic) cave houses. (p. 177)

Proceeding further, Peng and Song went on to discuss the situation for Tibetan minorities:

> Similarly, with the Tibetan tradition, distinct forms of mathematical systems are also evident… Examples include the prayer beads and abacus which are still used today as tools for calculation by many Tibetan people. Other tools specific to the Tibetan minority are wooden counting frames, which are used for calculating the time of the solar eclipse, lunar eclipse, and increscence. (p. 177)

Similar to the above examples, which dwell on the mathematics-related cultural practices among Chinese ethnic minorities, there are various mathematics-related activities, traditional everyday environmental practices, phenomena, and socially valued mathematical knowledge identifiable among Canadian Indigenous (Aboriginal) populations. Some of these, drawn from the Cree and Anishnaabe Aboriginal cultural groups, are shown in Table 1. Table 1 gives details of these identified traditional activities, environmental phenomena, everyday materials, and matches them with the mathematics concepts or topics to which they apply, based on the five mathematics strands of the Ontario (Canada) provincial mathematics curriculum, Grades 1–8 (2005).

6 Building Mathematics Education on Identified and Available Cultural Resources

Awareness has been created through research about the existence of mathematics-related traditional practices, everyday phenomena, and activities among ethnic minority and Indigenouscultures. The creation of this awareness amounts to surmounting the first of three hurdles usually encountered in developing a suitable curriculum, and then utilizing the developed curriculum to make mathematics education more meaningful, effective, and hence relevant for learners from these cultural backgrounds. The second hurdle centers on the effective integration of the compiled traditional cultural resources into a mathematics curriculum targeted on at-risk mathematics learners.

The advantages of such a curriculum are numerous. First, the curriculum would evolve naturally from, and build solidly on, the learners' lifeworld mathematical knowledge and experiences accumulated over time from their homes, the environment they live in, everyday activities, peer interactions, etc. From this solid mathematics premise acquired from their lifeworld, the learners would then transition to the subculture of school mathematics. If the mathematics curriculum is appropriately developed, such that this transition from the students' lifeworld mathematics to school mathematics is seamless, then a solid mathematics foundation would have been laid for the young learners from ethnic minority/Indigenous cultural backgrounds.

Aikenhead (1996) proposed and utilized the construct—cultural border crossing—to conceptualize the transition between the learners' lifeworld experiences and school learning. Applied to this particular

Table 1 Examples of traditional practices, environmental phenomena, everyday activities/materials (among the Cree and Anishnaabe) that apply to, and can be used for math teaching/learning (Ezeife, 2013a)

Traditional practice, everyday activities and materials, and phenomena	*Mathematics strand (Ontario curriculum) to which they correspond*	*Math topics or concepts where they (traditional practice, everyday activities, etc.) can be applied*
Flowers in the environment: While playing in the fields, young Aboriginal children would pick flowers and count the petals	Number sense and numeration; patterning and algebra	Numeric skills—counting, basic units of counting, different base systems. Naturalistic intelligence (Gardner's multiple intelligences); Patterns in the arrangement of the petals
The Spider Web—often seen in the usually rural environment of Aboriginal communities. The "Web" concept, the shape, and patterns in the "Web" relate to these traditional activities: Weaving of fabric, straw (for hats), and reeds (for traditional arrows and baskets)	Geometry and spatial sense; patterning and algebra	Geometrical shapes, areas of two-dimensional figures; operations on fractions—addition and subtraction; equivalent fractions and ratios
Ice holes and Ice Fishing	Measurement, data management and probability	Estimation/measurement of depth, probability of catching fish, volume of water in hole, problem solving, and reasoning/logic
Beadwork and beads strung and worn by the Anishnaabe	Patterning and algebra	Patterns in the beadwork, colors and ordering of beads
Traditional log cabins (called ***Os-ka-nsa***, in Cree language)	Geometry and spatial sense; Number sense and numeration	Width/length, areas, geometric figures, accounting—costs, money (counting/conversion)
The component groups of the "3 Fires Confederacy—(***Odawa***, ***Ojibwa***, and **Pottawatomie**)	Number sense and numeration	Set theory—set descriptions and symbols/notations, the universal set, complements and subsets

Traditional practice, everyday activities and materials, and phenomena	*Mathematics strand (Ontario curriculum) to which they correspond*	*Math topics or concepts where they (traditional practice, everyday activities, etc.) can be applied*
Making of moccasins. Traditionally, this usually involves measuring the feet of someone standing on the hide	Measurement	Units and standards of units, conversion between two different systems of units—the S.I. and the F.P.S. systems
The Bannock cake (called **Pah-ke-sikun** in Cree language) is a staple Aboriginal food	Geometry and spatial sense	Fractions, decimals, percents, concepts of symmetry and division
Traditional hunting using two types of arrowheads—round tips and sharpened tips. The round tipped arrows are used for hunting smaller game, while the sharpened tips are reserved for hunting larger game. Optimal hunting seasons/periods	Geometry and spatial sense; data management and probability. Number sense and numeration	Angles and shapes, speed and velocity of moving objects. Probability of catching hunted animals. Seasons of the year and their durations
Traditional housing—Lodges are usually shaped in a circular formation. The construction of the lodges is symbolic. At the center of the lodge is a hold for the fire, and at the top of the roof is a circle for smoke exit, while the doorways align with the four directions—East, West, North, and South.	Geometry and spatial sense	Coordinate geometry–directions and locations in space. The four cardinal points and formation of the four quadrants
Traditional fishing: The Anishnaabe make marsh grass in a circular pattern. A hole inside the circular formation is lined with tunnels for trapping fish. Often, six or more tunnels are linked to the hole	Data management and probability	Probability—its example and application in everyday life. (the Anishnabe technique involves running the fish through several tunnels until they are captured in one. This strategy adopts and exemplifies the principle of probability)

(continued)

Table 1 (continued)

Traditional practice, everyday activities and materials, and phenomena	*Mathematics strand (Ontario curriculum) to which they correspond*	*Math topics or concepts where they (traditional practice, everyday activities, etc.) can be applied*
Burial traditions: Burials are usually done on the 5th day after death, and the body is positioned to face the East. This tradition symbolizes a new beginning—the sun rises in the East	Number sense and numeration	The decimal (base 10) system of counting contrasted with the base 5 system, and base 2 system used in computer technology. The concept and use of the "place holder" in counting. Cycles and rotations. Directions—sunrise and sunset
Fish nets (called ***Anapi*** in Cree language), Canoes	Measurement; number sense and numeration	Principles of mass and weight, counting—number of fish a canoe can hold
Games and Sports—Running as a form of exercise, and in pursuit of hunted animals. Also, there are a lot of traditional Aboriginal games such as shell games, the campfire game, slide-of-hand tricks, the moccasin game, etc. In the moccasin game, for example, the target is for each player to correctly guess in which pouch a marked marble is hidden	Data management and probability, measurement	Principle of probability—Games involving chance, such as raffles, lotteries, and bingos. Determination of odds of winning. Measuring distances covered in a race, or during a hunting expedition

context, it means the transition from the students' lifeworld mathematics experience/knowledge to school mathematics experience. The seamless transition which is attained or accomplished if or when the students' lifeworld culture is in accord with, and supports the school mathematics culture, is referred to as smooth border crossing (Aikenhead & Jegede, 1999). This should be the desirable goal in the preparation of mathematics curricula in multicultural nations like Canada and China as they embark on all-inclusive school development programs for their citizens.

With an adequately developed mathematics curriculum in place, the third and final task to accomplish (that is, hurdle to surmount) is the effective classroom implementation of the developed curriculum. This goal will be attained if mathematics teaching and classroom learning experiences are selected such that they reflect the contents of the developed curriculum which, in turn, had taken into account, and built its foundation on, the lifeworld culture of ethnic minority and Indigenous learners. Apart from appropriate teaching/learning classroom experiences, there are other tangible factors to consider, such as teacher quality, quality of mathematics instructional practices (Cheng, 2014), and of course, an understanding by the teacher of the learning characteristics of the students being taught (Greenwood, de Leeuw, & Fraser, 2007). The issue of learning characteristics, which subsumes such contiguous concepts or factors as students' attitudes and habits, is an important issue with respect to the education of ethnic minority and indigenous learners. Research has shown that:

> High-context cultures are characterized by a holistic (top-down) approach to information processing in which meaning is "extracted" from the environment and the situation, [while] low-context cultures use a linear, sequential building block (bottom-up) approach to information processing in which meaning is "constructed". (Hollins, 1996, p. 134)

Since ethnic minority and Indigenous learners essentially belong to the high-context culture group (Hollins, 1996), their learning styles are strongly supported by the use of environmental traditional practices, phenomena, everyday activities, artifacts, stories, and familiar flora and fauna in mathematics teaching, as advocated in this chapter.

In pursuance of the current trend to build effective mathematics education for at-risk learners based on their culture, some researchers and educators in both the East (China) and the West (Canada) have developed teaching materials and produced well-researched textbooks geared toward meeting the specific needs of these learners. During his recent (2014) research field trip to Southwest University (SWU) in Chongqing, China, under the auspices of the University of Windsor-Southwest University Exchange Program, Ezeife interacted with mathematics educators and researchers in SWU, and mathematics teachers in several affiliated schools in the city of Chongqing and its environs. In the course of these highly productive interactions, mathematics education, especially for at-risk learners, was extensively discussed, and a new book series for mathematics teaching at the elementary school level became the focal point of discourse (Ezeife, 2014).

The series, aptly entitled, *Elementary mathematics-culture book series*, edited by Song and Zhang (2014), contain ten subvolumes (books) with the respective titles (translated into English from the original Chinese language), thus:

> Nature and Mathematics, Games and Mathematics, Environment and Mathematics, Life and Mathematics, Health and Mathematics, Science and Mathematics, Economics and Mathematics, History and Mathematics, Art and Mathematics, Mathematicians and Mathematics.

The appeal of the mathematics-culture book series is that the series drew on the day-to-day life experiences and the environment of mathematics learners and used these experiences as building blocks on which mathematics education is firmly anchored. This is the approach advocated by contemporary researchers and mathematics educators for eliminating the lacuna or chasm—often perceived by ethnic minority and Indigenous students between their lifeworld and the world of school mathematics, a situation that usually puts them at risk in their study of mathematics.

Some detailed contents of the mathematics-culture book series from two subvolumes of the series, namely "Life and Mathematics" and "Games and Mathematics," are shown in the first column in Appendix I, while the researcher's (Ezeife's) comments on the concepts and the areas of mathematics to which the specified contents could be applied in actual classroom teaching situations, and the cultural, environmental mathematics knowledge inherent in, and discernible from the contents, are also given (shown in the second column, in Appendix I). In Appendix II, the researcher provides a sample lesson suited to the environmental outdoor life in typically rural Indigenous communities in Canada. The lesson notes and learning experiences dwell on activities specifically designed to engage the students in see-mathematics-in-your-environment, learn-as-you-do sessions. Appendix III shows the use of the "Spider Web" concept (which is relatable to the concept of *Partitioning* in mathematics) in teaching Operations on Fractions, Areas of two-dimensional figures, etc., as earlier specified in Table 1.

7 Next Phase

This chapter has engaged in a research-based discourse of the status quo in contemporary mathematics education for at-risk mathematics learners in China (ethnic minority students) and Canada (Indigenous/Aboriginal students). The chapter postulated that these ethnic minority and Indigenous students will be more attracted to the study of mathematics and fare better in examinations, if mathematics is presented to them as part and parcel of their lifeworld, by using their culture, environment, mathematics-related activities/practices, and ways of knowing, to teach them. The chapter went further to present and discuss some curriculum development efforts that have been made by current mathematics educators/researchers, and the culture-oriented teaching/learning materials and texts developed by these researchers.

Since these materials are widely available for use, it is hoped and expected that future studies and projects of this nature—based on the "Cultural Perspectives" goal of the Mathematics Research Team of the Reciprocal Learning Project—will engage in extensive fieldwork in both Canada and China to try out the efficacy or effectiveness of the developed materials in an experimental/empirical field setting. "Sister Schools" (Connelly & Xu, 2014) and available ethnic minority/Indigenous mathematics learners from both countries (China and Canada) would be the feasible targets and would hopefully be fully involved in this anticipated follow-up work.

8 Summary, Suggestions, and Conclusion

This chapter drew attention to the yearning need for mathematics education in general, and in particular, meaningful and relevant education for ethnic minority and Indigenous learners of mathematics. It is strongly suggested that, in School Development Programs in all climes and societies, mathematics education should be embarked upon with the infusion of a high dose of cultural content. This is in recognition of the visible and tangible role culture plays, not just in relevant and effective mathematics learning in schools, but also in ensuring the establishment of all-inclusive, equitably developed and implemented school programs for all students, irrespective of their cultural backgrounds (Dei, 2015).

It is further suggested that school programs, especially mathematics education programs, be developed and implemented in such a way that schools continue to sustain their age-old role of overall societal development and regeneration. In making this suggestion, this chapter took cognizance of the fact that such revered structural edifices as the *CN Tower* in Canada and the world-famous *Great Wall* in China are mathematical, architectural, and technological wonders which owe their emergence to deeply rooted ingenuity, creative school development, and other contiguous factors in East–West societal civilizations that serve as shining pointers to, and strong indicators of, the ever-growing cultural role which mathematics plays in modern-day global human existence.

By harping on the need for the utilization of culture and adopting culture-sourced content and procedures that would make mathematics relatable, meaningful, and hence, relevant—as suggested in this chapter, it is projected that students' attitudes toward mathematics, particularly at-risk learners, would change for the better. The culture-sourced content and procedures may actually be drawn from the liberal arts because as research has revealed, a firm grounding in the liberal arts could "give shape and humanistic substance to mathematics and science education" (Kotsopoulos, 2015, p. 35). The "shape and humanistic substance" may serve as possible catalysts that could help alter students' prevalent negative attitudes toward mathematics. This profoundly negative perception of mathematics that has, unfortunately, persisted over the years is reflected in the statement by the National Research Council (NRC, 1989), in these words:

> Public attitudes about mathematics are shaped primarily by adults' childhood school experiences. Consequently, mathematics is seen not as something that people actually use, but as a best forgotten (and often painful) requirement of school. For most members of the public, their lasting memories of school mathematics are unpleasant – since so often the last mathematics course they took convinced them to take no more. (p. 10)

It is hoped that this study, along with other research engagements undertaken by the Mathematics Research Team of the Reciprocal Learning Project (RLP) between Canada and China, will contribute significantly to the improvement of mathematics education, more so for at-risk learners from ethnic minority and Indigenous cultural backgrounds in both countries—Canada and China—within the seven-year duration of the Reciprocal Learning Project.

Acknowledgements Dr. Anthony N. Ezeife gratefully acknowledges the "permission to reprint" that was granted by the Editorial Board of the Comparative and International Education Journal (CIE, Vol. 45, Issue 1, 2016) in which the article had been originally published.

Appendix I

See Tables 2 and 3.

Appendix II

A lesson on mathematics in the community, entitled, Outdoor Mathematics [Mathematics in the Park] (Ezeife, 2013b)

A schedule of timed mathematics learners' activities (Targeted to Grades 5–8 Ethnic minority/Aboriginal students).

1. **Activity 1: Look and Write** (15 minutes)

Study (look closely at) this field/park and write down whatever you see in the park that reminds you (tells you) about any part of mathematics you have learned in your class.

<u>Note</u>: Write this in the small notebook given to you for outdoor mathematics activities.

Activity 2: Count and Record (10 minutes)

Table 2 Elementary mathematics-culture book series: Life and Mathematics sub-volume (Song & Zhang, 2014)

Some contents of the book (translated from the original Chinese language)	*Researcher's comments on (and insight into) the mathematics concepts/areas to which the contents could apply; and their cultural significance in mathematics teaching*
Calculating (managing) your time before going to school	This suggests 'waking up with mathematics'- mathematics (the clock) as the day's guide—learner's culture. Start of the child's day and preparation for school are typical day-to-day activities of the child. Thus, the child sees mathematics as part and parcel of his/her day, and hence, an integral part of life
Time management on your way to school	Here, again mathematics is the day's guide. Thus, the child is 'thinking mathematics' as s/he walks to school
How to use "Coordinates" in mathematics to get a seat in class	Coordinates (hence, Coordinate Geometry/Cartesian System). This is a direct application of the mathematical knowledge that the student has acquired
Nutrition facts about the food in your (student's) lunch box	Calculating (and thus, knowing the components and percentages of the food items in a typical student's lunch box/snack box). This deals with the important topic—nutrition facts and their bearing on the health/life of a student
The patterns and shapes on the blackboard	Patterns and shapes on the class blackboard may likely relate to mathematics teaching/learning tools and manipulatives.
Calculations in a farm	Farm work, which will most likely be one of a typical student's after-school activities/engagements (chores) would entail a lot of mathematics-associated exercises. For instance, the child may be involved in picking (and most likely, counting) apples, tomatoes, etc.; plucking oranges, and pears and heaping them into baskets or containers for weighing. Thus, the child learns the use of scales, weights and measures (in mathematics) as s/he engages in typical everyday chores and activities.
Doing the laundry	The concept of counting—Number (and types) of clothes washed. Weight of clothes. Volume of water used

Some contents of the book (translated from the original Chinese language)	*Researcher's comments on (and insight into) the mathematics concepts/areas to which the contents could apply; and their cultural significance in mathematics teaching*
Using knowledge about angles in playing soccer	This concerns the mathematics in school games like soccer, hockey, tennis, ping-pong, and other popular games students play. Angles and trajectories could be discussed in this context
A birthday gift for Dad	Mathematics in family life: Birthday gifts for family members; sizes of clothes, shoes, and hence the concept of measurement. Wrist sizes, circumferences of the head, arms spans, etc. of family members could be measured
The mathematics in toys	Young mathematics students should be guided to look at every toy they use as a mathematical contraption (device). This should naturally lead them to ask the question: Where is the mathematics in this toy?
How to draw a map	Scales and measures, distances, plans and elevations could be covered effectively under this topic, making effort to relate these to the day-to-day life of the students
Shopping with your Mom	The growing child comes in contact with money at some stage. This should be a good opportunity to relate money to mathematics, by discussing money denominations and their relative values; and thus, their ratios with respect to one another
The "magic" mirror.	Generally, young students love mirrors and mirror games. This situation can be beneficially tapped when mirrors are used to teach the mathematics concepts of reflection, angles associated with reflection, distances of objects and images to/from mirrors, and the concept of lateral inversion. The kaleidoscope (a popular children's toy) could be shown in class/discussed in relation to its mathematics content
The mathematics in climbing a mountain	Mountain climbing is usually an enjoyable outdoor activity for students. The mathematics in this exercise—tangents and slopes, angles, curves, tracks, and elevations, should be harped upon as/when children engage in the exercise/activity

Table 3 Elementary mathematics-culture book series: Games and Mathematics sub-volume (Song & Zhang, 2014)

Some contents of the book (translated from the original Chinese language)	*Researcher's comments on (and insight into) the mathematics concepts/areas to which the contents could apply; and their cultural significance in mathematics teaching*
The mathematics in planting trees	Applicable mathematics teaching concepts include land areas and distances, map drawing, spaces between trees. The knowledge acquired while studying this topic in a mathematics class early in life could become useful in a future farming career or landscaping
The Chinese paper cut	'Paper cutting' is a cultural play activity that has a good deal of mathematical implications
The mathematics in "clocks"	The clock should be presented as a commonly used cultural instrument in which a lot of mathematics can be seen. The use of the clock in modern time-telling should be contrasted with the traditional time-estimation in ancient times (for example, by observing the varying lengths of shadows)
How much juice can you have?	Students encounter fluids and liquids in their environment on a daily basis. The mathematics of 'volumes' would come up naturally in class as liquids and the containers/vessels that hold them are discussed
The mathematics in gambling	Gambling is a common activity in many cultures which students will encounter when they come of age. Mathematics should be used to reveal the odds of winning/losing under the concepts of Permutations and Combinations. Problem gambling and addiction should be mentioned and discouraged
The "Bumper Car" and numbers	Various number games, usually played by children in table formats, such as mathematical games involving the "addition truth table" and "multiplication truth table" fall into the "Bumper Car" category

How many trees are there in this park?

Activity 3: Observe and Sort (20 minutes)

3(a) In this activity, you will put the trees you counted in three categories or groups, according to their size—small, medium, large.

i. How many of the trees are small in size?
ii. How many trees are of medium size?
iii. How many trees are large?

3(b) If you are asked to draw a graph to show the number of trees in the park based on their sizes (small, medium, large), what type of graph will you draw?

Note: Choose **one** type of graph from the list below:

1. Line graph
2. Straight line graph
3. Stem-and-Leaf plot
4. Bar graph
5. Pictograph

Activity 4 (Group Activity): Measure and Record (25 minutes)

In this activity, you will work in pairs, that is, each person will work with a partner.

4(a) Choose any three trees in the park such that one tree is small, one is medium, and one is large.

4(b) Using the tape given to your group, measure the circumference of (distance around) each of the three trees, and record your results:

i. The circumference of the small tree is?
ii. The circumference of the medium tree is?
iii. The circumference of the large tree is?

4(c) Measure the diameter of each of the three trees and from your measurements, calculate (find) the radius of each type of tree, and record your results:

i. What is the radius of the small tree?

ii. What is the radius of the medium tree?
iii. What is the radius of the large tree?

Activity 5: How many trees (2 minutes)

If you are taken to a park two times the size of this park, about how many trees do you think that park will have? (Assume that trees in all parks have the same spacing, that is, the same distances from one tree to another, just as the trees in this Particular Park in which you did today's outdoor mathematics session).

Activity 6: General Comments (10 minutes)

5(a) Did you enjoy today's mathematics class that you did outdoors (that is, outside your regular classroom)?

5(b) Give a reason for your answer. That is, state why you enjoyed or did not enjoy today's outdoor class.

5(c) Do you think mathematics is part of your daily life, or is it just something you do in the classroom?

5(d) From now on, will you try to think of examples of mathematics in your environment or community:

i. When you are going to school? Yes or No ---------
ii. When you are going back to your house after school? Yes or No ----------
iii. When you are at home? Yes or No ---------

Activity 7: Project Work

7(a) **Estimating the ages of trees in the park (For Grade 8 students only).**

Based on your measurements of the diameters of the trees in Questions 4(b) and 4(c) for the small, medium, and large trees, estimate:

I. The age of the medium tree (that is, how old is the medium-sized tree?)
II. The age of the large tree.

[<u>**Hint/Clue**</u>: Assume the small tree in Questions 4(b) and 4(c) is 4 years old. Then use the **ratio** approach (proportions) to find (estimate) the ages of the medium and large trees. Thus, you have to form ratio equations (involving the diameter of the small and medium trees first, and after that, the diameters of the small and large trees) to enable you do the estimation].

7(b): **Designing an Aboriginal-oriented park**

As a hands-on, learn-as-you-do project exercise for this outdoor mathematics class (Mathematics in the Park), design your ideal Aboriginal park that should take into consideration some specific design issues/characteristics, and which will include the following items:

I. The **spacing** (that is, the distances) between the trees in the park. These distances should be labeled in your completed design.
II. The ideal **number** of trees in the park.
III. The **mix** of trees in the park (that is, the number of small, medium, and large trees).
IV. The **type** of trees in the park. Here, you should think of trees that reflect the Aboriginal culture, and are traditionally planted and used in Aboriginal societies and communities in Canada.
V. An **orchard** (a fruit garden) in the park. Here, think of the type of fruits that are symbolic of, and important to Indigenous peoples of Canada. (You can ask your parents, guardians, older siblings, and community Elders to give you some ideas and hints about such traditionally meaningful trees).

Note: We hope you enjoyed today's outdoor mathematics class which has the goal of bringing out clearly the fact that mathematics is part and parcel of your everyday life, and has been done and practised by Aboriginal people, and other Indigenous cultures, for many centuries.

Note: This outdoor mathematics class (Ezeife, 2013b) was designed and utilized for teaching Indigenous (Aboriginal) students (Grades 5–8) in the University of Windsor's 4-Winds STEM (Science, Technology, Engineering, and Mathematics) Project, 2012–2014. The project was launched to attract Aboriginal students in Windsor and its environs to STEM fields of study by presenting course contents to them using culture-oriented and environmentally sourced materials from their life-world.

Appendix III

Teaching mathematical "Operations on Fractions", Equivalent Fractions, and Areas of two-dimensional figures using the 'Spider Web' concept

Example problem: Find 3/4 + 5/6

Solution using the 'Spider Web' concept: 3/4 + 5/6 = 18 rectangles + 20 rectangles (obtained by directly counting the number of rectangles that make up the fractions 3/4—three quarters, and 5/6—five sixths, respectively, in Fig. 1)

= 38 rectangles

But, the total number of rectangles in the Spider's Web

= 24 (6 columns x 4 rows)

Thus, 3/4 + 5/6 = 38 rectangles

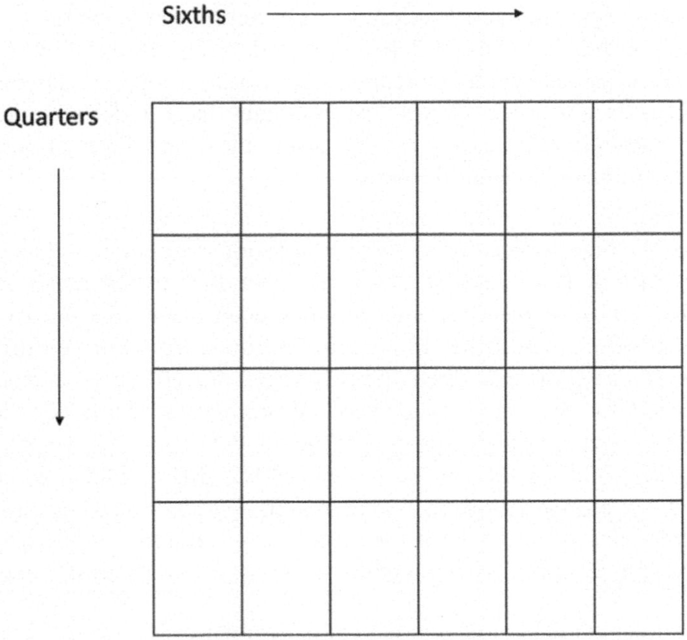

Fig. 1 The Spider's Web

Now, when this number (38) is compared to the total number of rectangles (24) in the Spider's Web, then we see that this implies $38/24 = 19/12 = 1\ ^{7}/12$

Hence, $3/4 + 5/6 = 1\ ^{7}/12$

Follow-up exercise and consolidation: Find $1/4 + 1/6$, using this method (the 'Spider Web' concept)

References

Aikenhead, G. S. (1996). Science education: Border crossing into the subculture of science. *Studies in Education, 27,* 1–52.

Aikenhead, G. S., & Jegede, O. J. (1999). *Cross-cultural science education: A cognitive explanation of a cultural phenomenon.* Retrieved from http://www.usask.ca/education/people/aikenhead/970759.htm.

Arora, A., Foy, P., Martin, M. O., & Mullis, I. V. S. (2012). *TIMSS 2011 international results in mathematics.* TIMSS & PIRLS International Study Center, Lynch School of Education, Boston College, Chestnut Hill, MA, USA and International Association for the Evaluation of Educational Achievement, IEA Secretariat, Amsterdam, The Netherlands.

Ascher, M. (1998). *Ethnomathematics: A multicultural view of mathematical ideas.* Boca Raton, FL: Chapman & Hall/CRC Press.

Binda, K. P. (2001). Aboriginal education in comparative and global perspectives: What has research and practice done for Aboriginal education in Canada? *Canadian and International Education, 30*(1), 1–16.

Bourque, J., Bouchamma, Y., & Larose, F. (2010). Aboriginal students' achievement in science education: The effect of teaching methods. *The Alberta Journal of Educational Research, 56*(1), 57–71.

Center for Effective Collaboration and Practice (CECP). (2001). *School development program.* Retrieved from http://cecp.air.org/resources/success/school_development_program.asp.

Cheng, Q. (2014). Quality mathematics instructional practices contributing to student achievements in five high-achieving Asian education systems: An analysis using TIMSS 2011 data. *Frontiers of Education in China, 9*(4), 493–518.

Connelly, M., & Xu, S. (2014). Sister school pairings among Canadian and Chinese primary and secondary schools in the Canada-China Reciprocal Learning Project (RLP). *Envisioning Reciprocal Learning between Canada and China.* Retrieved from http://pc240-12.eduf.uwindsor.ca/Can-Cn/index.html.

Dei, G. J. S. (2015). Reflections on "dropping out" of school: Meeting the challenge of youth engagement. *Education Canada, 55*(2), 13–17.

Ezeife, A. N. (2013a, November 4–6). *Teaching mathematics with an Indigenous flavour: The implementation of a culture-oriented mathematics education program for Canadian Aboriginal youth.* Paper presented at the London International Conference on Education (LICE), London, UK.

Ezeife, A. N. (2013b). *Outdoor mathematics: Math in the Park.* Lesson notes and activities designed for Aboriginal students (Grades 5–8) in the University of Windsor Science, Technology, Engineering, and Mathematics (STEM) Project, funded by the University of Windsor Strategic Priority Fund (SPF), 2012–2014.

Ezeife, A. N. (2014). *A lived educational experience: Interactive and reciprocal.* A report on the educational field trip (May 4–June 5, 2014) made to China under the auspices of the Southwest University (Chongqing, China)–University of Windsor (Ontario, Canada) Exchange Program.

Friesen, J. B., & Ezeife, A. N. (2009). Making science assessment culturally valid for Aboriginal students. *Canadian Journal of Native Education, 32*(2), 24–37.

Gaskell, J. (2003). Engaging science education within diverse cultures. *Curriculum Inquiry, 33*(3), 235–249.

Greenwood, M., de Leeuw, S., & Fraser, T. N. (2007). Aboriginal children and early childhood development and education in Canada: Linking the past and the present to the future. *Canadian Journal of Native Education, 30*(1), 5–18.

Hollins, E. R. (1996). *Culture in school learning: Revealing the deep meaning.* Mahwah, NJ: Lawrence Erlbaum Associates.

Kotsopoulos, D. (2015). How the liberal arts made me better at mathematics. University Affairs, *56*(6). Retrieved from www.universityaffairs.ca/june-july2015.

Luangrath, P., & Vilaythong, T. (2010). An analysis of students' perceptions of physics in science foundation studies at the National University of Laos. *Canadian and International Education, 39*(1), 32–40.

Moreno-Garcia, S. (2012, December 11). Cultural barriers crunched by numbers. *UBC Reports.* Retrieved from http://aboriginal.ubc.ca/2012/12/11.

National Council of Teachers of Mathematics. (2000). *Principles and standards for school mathematics.* Reston, VA: Author.

National Research Council. (1989). *Everybody counts: A report to the Nation on the future of mathematics education.* Washington, DC: National Academy Press.

Peng, A., & Song, N. (2014). The status quo development of mathematics education for ethnic minorities in China. *Studies in Ethnicity and Nationalism, 14*(1), 171–187.

Rong, M. (2006). *Education of ethnic minorities in contemporary China*. Retrieved from www.case.edu/affil/tibet/more.Tibetinfo/documents.

Sicat, L. V., & David, M. D. (2011). Performance in basic mathematics of indigenous students. *2011 International Conference on Social Science and Humanity. IPEDR, 5*. Retrieved from http://www.ipedr.com/vol5/no2/75-H10196.pdf.

Song, N., & Zhang, Y. (2014). *Elementary mathematics-culture book series*. Chongqing: Southwest Normal University.

Snyder, T. D., & Dillow, S. A. (2011). *Digest of education statistics 2010. NCES 2011-015*. Washington, DC: National Center for Education Statistics.

The Ontario Curriculum, Grades 1–8: Mathematics (2005). Toronto: Ministry of Education.

Thomson, S. (2009). International mathematics and science study shows mixed results for Australia. *Research Developments, 20*, 1–4. Retrieved from http://research.acer.edu.au/resdev/vol/20/iss20/13.

Warren, E., Baturo, A., & Cooper, T. (2004). Queensland Indigenous Students in rural and remote classrooms: Pedagogical practices and numeracy learning. In J. Zajda & P. Ninnes (Eds.), *Proceedings of the 32nd Annual Conference of the Australian and New Zealand Comparative and International Education Society: Global Pedagogies: Equity, Access & Democracy in Education* (pp. 153–163), Melbourne, Victoria. Retrieved from https://www.researchgate.net/publication/27468894_Queensland.

Index

S. C. Zhu et al. (eds.), *Reciprocal Learning for Cross-Cultural Mathematics Education*, Intercultural Reciprocal Learning in Chinese and Western Education,
https://doi.org/10.1007/978-3-030-56838-2

The manufacturer's authorised representative in the EU is Springer Nature Customer Service Centre GmbH, Europaplatz 3, 69115 Heidelberg, Germany. If you have any concerns regarding our products, please contact ProductSafety@springernature.com

Printed and bound by CPI Group (UK) Ltd, Croydon, CR0 4YY
30/09/2026
02225443-0009